The Archaeology of Water Supply

Edited by

Marta Żuchowska

BAR International Series 2414
2012

Published in 2016 by
BAR Publishing, Oxford

BAR International Series 2414

The Archaeology of Water Supply

ISBN 978 1 4073 1012 1

BAR Publishing is the trading name of British Archaeological Reports (Oxford) Ltd.
British Archaeological Reports was first incorporated in 1974 to publish the BAR
Series, International and British. In 1992 Hadrian Books Ltd became part of the BAR
group. This volume was originally published by Archaeopress in conjunction with
British Archaeological Reports (Oxford) Ltd / Hadrian Books Ltd, the Series principal
publisher, in 2012. This present volume is published by BAR Publishing, 2016.

Printed in England

BAR
PUBLISHING

BAR titles are available from:

BAR Publishing
122 Banbury Rd, Oxford, OX2 7BP, UK
EMAIL   info@barpublishing.com
PHONE   +44 (0)1865 310431
FAX   +44 (0)1865 316916
www.barpublishing.com

CONTENTS

DRAINAGE SYSTEM OF THE RAINWATER AND THE EXCESS WATER DISCHARGED ON THE STREETS OF POMPEII

Yoshiki HORI
Kyushu University

Key Words: water drainage, run-off, *lanfricariae*

Introduction

In Pompeii, the fact that water was evacuated to the streets from drains is attested by the many outlet holes in the kerbs outside the entrances. While isolated drainage and cisterns have been identified in many houses, Pompeii did not have a general underground drainage system covering the whole city (Figure 1).[1] The present incline of the ground in a north-south direction, exceeding that in the east-west direction, is 30m/800m (3.75%, Figure 2).[2] Unless the surface could be levelled to make a flatter surface, Pompeiian builders could not construct a main conduit running from north to south. (It was necessary, or at least recommended, that the declivity of the main conduit should be no more than 2.0% degrees. Thus the builders could avoid sweeping the blocks and injuring the beds or side walls by any heavy discharge of rainwater shortly afterwards.)[3] As M. Koga has observed, the rainwater flowed onto the via Stabiana and partially into an underground drain. Then it flowed to the north-south sewer under the middle of the *insulae* west of the main axis, under the Large Theatre, to the portico of the Large Theatre to the south, where it was discharged beyond the city wall (Figure 3) (See KOGA, 1992, 57-72). The ground level to the south of the Large Theatre probably corresponds to the original surface of the *insulae*. Its level is about 10 metres lower than the underground drain running along the Triangular Forum, thus avoiding a steep slope on which the water would run fast.

A steeper incline that is not as critical as in free-flow aqueduct channels will not increase the volume of discharge (output of water equals input, regardless of speed), but it will accelerate the flow, which will run faster in the underground channel. Since it is never easy to keep drains clean, and probably the drains would become easily and completely blocked with filth and debris from the street, it is difficult to give regular maintenance to a drain on a steep gradient.

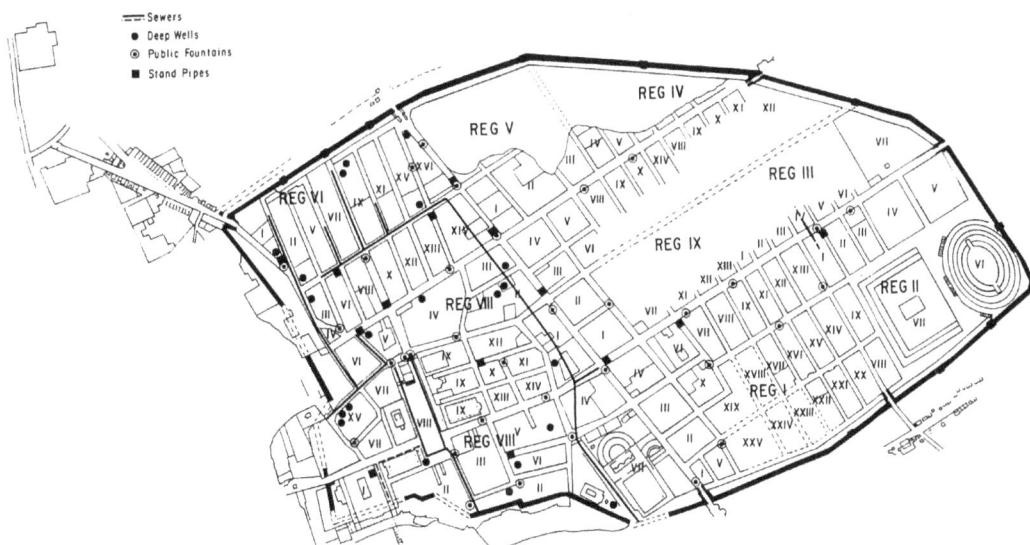

Fig.1. Water supply and drainage system in Pompeii (after RICHARDSON,1988)

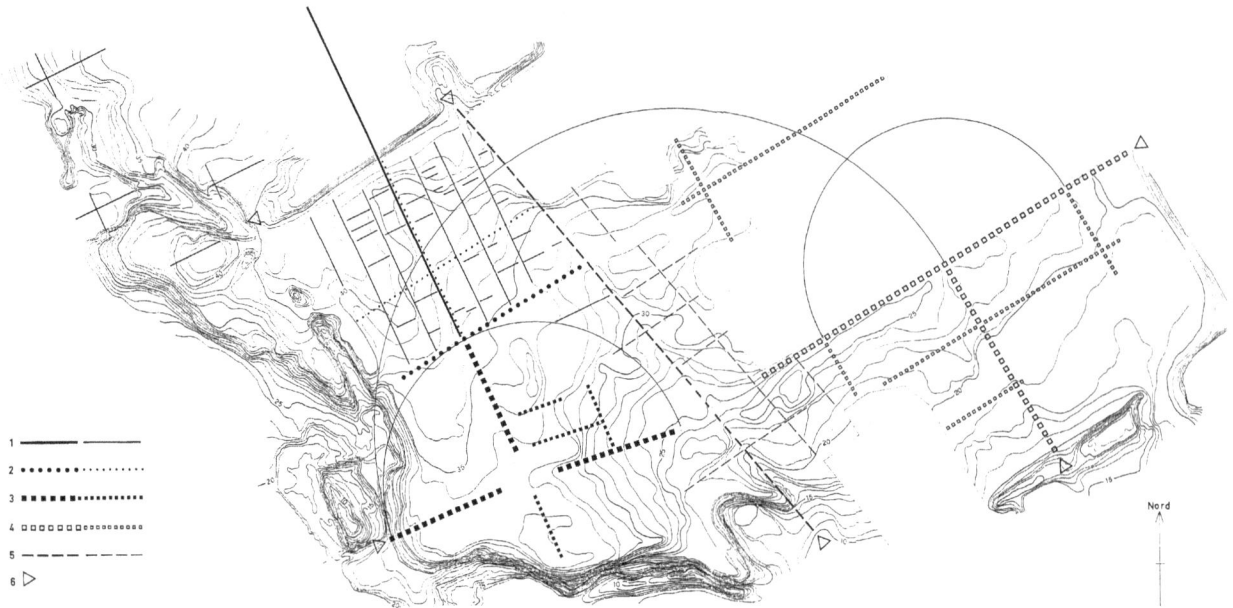

Fig. 2. Ground level of Pompeii (after SOMMELLA,1995)

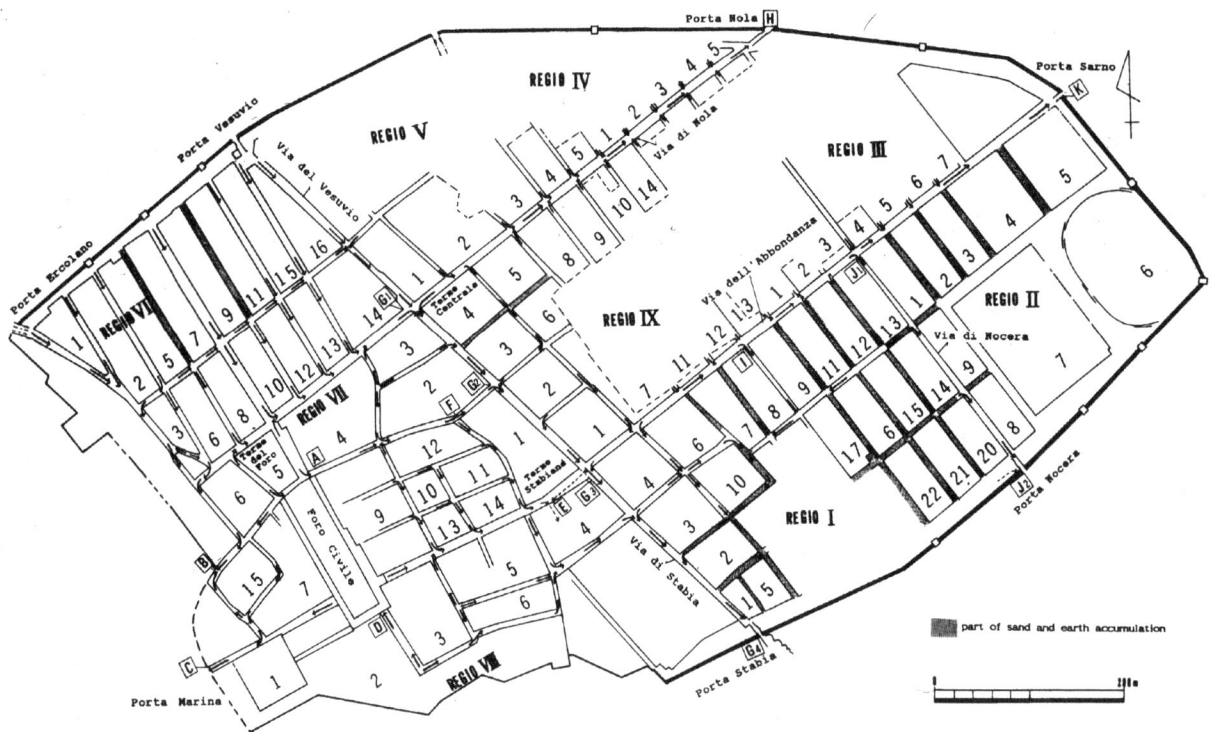

Fig. 3. Surface drainage system of Pompeii (after KOGA,1992)

The construction of a free-fall drainage conduit running from north to south would have been possible only if 12-metre-deep catch basins could be provided every 400m (or, respectively, 3m every 100m). However, the construction of such basins was technically and financially impossible. Pompeiian builders would have been forced to dig into the hard lava bed rock. Only on via Stabianna would they have to dig more than 10 basins 3m deep. We have seen that the largest of Pompeiian sewers,

the Holconius crossing - originally a stream running from the Forum through the via dell'Abbondanza - collected a part of the drain water from the Stabian Bath. Surface water that collected from the north and west quarter of Pompeii was conducted (Figure 4), whenever possible, through the Stabian gateway, along the street and, where possible, in a covered conduit. Its incline (more than 4.7%, even 8.4% in the southern part beyond the via del Tempio d'Iside) could be determined on the basis of the difference between the ground levels on the Holconius crossing and the uncovered drain below the street at the south end of the via Stabiana. Basins would have been necessary if running water could damage the blocks and beds of the conduit.

While the ground floors of Pompeiian houses corresponded to the general incline of the ground (Figures 1, 2 and 3), in the limited area of the Forum in the west and the Palestra in the east the drainage system was able to drain water despite minor alterations in the landscape. It would not be possible on the major ground elevation.

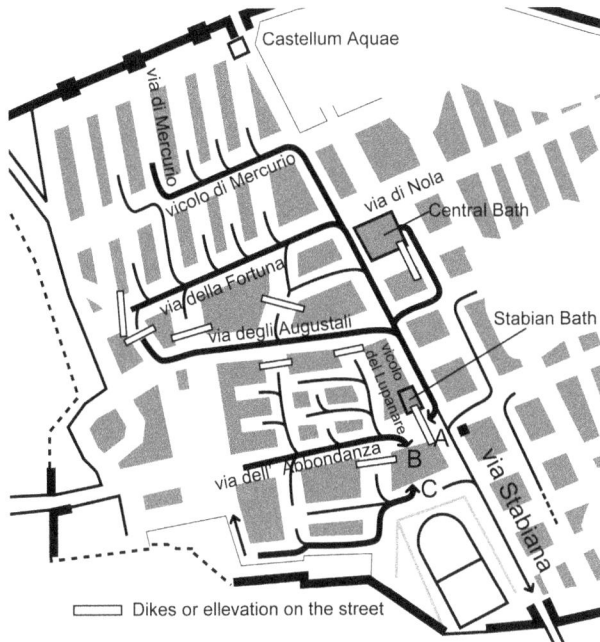

Fig. 4. Water flow on the streets

The lack of an underground urban drainage system covering the whole city of Pompeii contrasts strongly with the equivalently efficient system of the water supply. The function of its water towers was to receive water from the preceding section at a similar height, and to deliver it from this same height into the next section, thereby from the Castellum Acquae into three sections, the total distance to be covered under pressure. There must have been an open-air tank at the top of each tower, into and out of which the water flowed at ordinary atmospheric pressures. Any excess water supplied to Pompeii was finally discharged from public fountains to the streets (Figure 1).[4]

A drainage system covering the whole city was practically desirable as well, making this function far more important than any minor civil engineering role the independent cisterns and drainages may have had. The north-south axial street (the via Stabiana) was designed to run more or less along the valley between two ranges of hills to the east and west of Pompeii (Figure 2). The natural ground level within the city of Pompeii sloped steeply downwards from Vesuvius to the Stabiae Gates: a drop of nearly 35m. Therefore an additional function of the running water on the road surface was, no doubt, to clean the filth from the street. Yet it is impossible to determine whether the Pompeiian builders planned the street to become public

drainage, or whether it was that the sewers were open on the main streets, including via dell'Abbondanza and via del Tempio di Iside. Surely the main purpose was to control the water running on the steep road surface, corresponding to the original ground level.

Laser Scanning in Pompeii

Over two seasons (from August to September in 2006 and from December 2006 to January 2007), a topographical investigation of Pompeiian streets was carried out (Figure 5).[5] For the last three decades we have depended on the famous and detailed topographical research by H. Eschenbach (1979), which is, however, a prime example of a more sporadic and limited style of investigation.

In his pioneering map of running water on the Pompeiian streets, published in 1992, M. Koga observed that the rainwater running on the surface of Pompeiian streets was carefully controlled, and that street surfaces were constructed according to a detailed design. This is a perfect example of a recent field study emphasising a secondary function of streets, in this case acting as a channel. Such secondary functions used to be neglected or poorly documented by previous research. The investigation in 2006 and 2007 has confirmed the substantial accuracy of Koga's observations.

In this investigation we introduced the most advanced measuring method: a 3D scanning system, in which more than 10,000 scanning laser beams can be emitted in one second and the object can be described as clouds of points having three-dimensional coordinates. The measuring of all the streets was completed within two seasons. To cover the whole city, the network of streets was separately measured into more than 300 clouds of points. The main streets, such as via Stabia, via dell'Abbondanza and the eastern stretch of via di Nola connecting the four gates of Pompeii (Vesuvius, Nola, Sarno and the Stabian Gates) were also measured in order to map their three-dimensional positions on an existing map and to locate the water towers and public fountains.

Fig. 5. Scanned area (point clouds)

Since each 3D model derived from that measuring system has local coordinates, it is necessary to merge point clouds and convert their coordinates to global ones (longitude and latitude). And some markers of which the longitude and latitude were measured by Arch Rispoli in 2006 were included into our scanning data in order to locate our data on the GPS (Global Positioning System), which could provide the first 3D model of the whole of the street network of Pompeii. Some errors that seem to have occurred in the coordinates of the merged clouds can be attributed simply to dimensional inaccuracy in laser scanning, but are within the accepted 0.05% error rate (see Table).

Table: Errors in measurements (metres)

Points	Results from Rispoli			Results from Laser Scanning			errors in measuremets		
	x	y	z	x	y	z	x	y	z
1	0.000	0.000	41.828	0.179	0.172	41.812	0.1.9	-J..7J	0.016
2	625.545	-167.561	25.845	625.521	-167.468	24.623	0.024	-J.09J	1.222
3	541.389	-72.271	30.319	541.362	-72.215	29.800	0.027	-0.055	0.520
4	973.676	-175.092	15.558	973.740	-175.088	16.629	0.0C4	-J.004	?.07?
5	880.918	-21.603	22.955	880.975	-21.574	22.019	-0.057	-0.029	0.936
6	194.429	-145.896	37.510	194.650	-145.685	37.626	0.221	-0.211	-0.11C
7	239.643	165.814	47.212	239.568	166.196	48.064	0.075	-0.382	-0.852
8	569.703	-203.700	24.527	569.712	-203.732	23.491	-0.009	0.032	1.036
9	443.598	-258.308	27.129	443.543	-258.214	26.467	0.055	-0.091	0.662
10	430.107	-366.258	32.253	430.564	-366.102	31.388	-0.457	-0.150	0.865
11	268.594	-177.884	35.029	268.476	-177.336	35.018	0.118	-0.548	0.011
12	434.978	-138.525	29.951	434.942	-138.489	29.535	0.036	-0.037	0.416

Pompeiian streets and running water on the surfaces

Pompeii was designed in the 6th century BC, and it is not clear how the streets, which were repaired and raised later, were originally laid out. At that time, the Greeks in Sicily and the Etruscans had started to plan their new towns in a rational and regular way. In archaic Etruria, the new systematisation of town planning resulted from the omnipresent Greek influence, in particular from the ideas formulated by Hippodamus of Miletus in the 5th century BC. The regular *insulae* at Capua received their final shape at the time of the large extension when Capua received the Samnite immigrants in 423 BC, according to Livy (VI, 37).

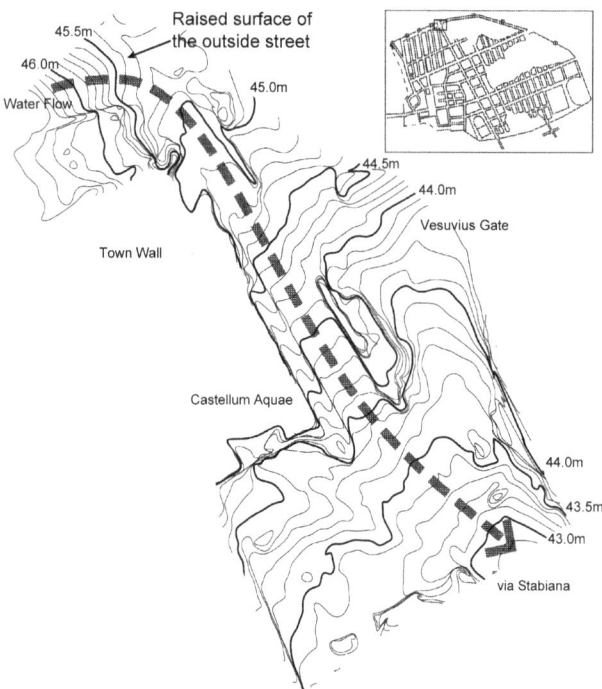

Fig. 6. Raised ground of the paved surface, and the water flow

The first major development occurred probably in the 3rd century BC. The framework of main streets became the point of origin for the construction of three main component parts of the street plan. This fact indicates that the street layout in different parts of Pompeii was established at the same period. Although differences in the alignment of streets not crossing at right angles could be related to instances of town planning, its early phase, the terracing down a shelving valley along the via Stabiana, seems to have been shaped by the topographical and geographical convenience of drainage of the run-off flowing on the streets, rather than by any coherent plan. Remodelling by the creation of a rectangular Forum, under Augustus, did not lead to a reshaping of the whole town (DOBBINS, 1997, 73-87). The most we can say is that there is a possibility, but no more, that via Stabiana still functioned as a main drain line in the final phase.

At the same time, the streets within that framework could be repaved and raised as well, and they could be systematically repaired. Outside the Nuceria and Vesuvius Gates, pebbled roads datable to the late 3rd century BC have been excavated beneath the later road surfaces, which were raised (Figure 6) (NAPPO, 1997, 91-120). The town wall was sunk through alluvium or garbage disposal to a depth of approx. 1.5m, deposited by rainfall running along the spine of Mount Vesuvius. Hence the Vesuvius Gate lay below the road running along the northern town wall, and the run-off would be gathered into that gate. The run-off coming from the mountain after rain could flush down the garbage on the main traffic route of Pompeii, when a given volume of the water ran fast, but also the debris of earth from the run-off could accumulate on the street. Pompeiian engineers probably needed to ensure that the gate collecting the run-off would be adequate to handle the total volume of water.

Contrary to the Vesuvius Gate, the street level at the Herculaneum Gate was higher than the ground level of the town wall. This prevented the run-off from flowing through the gate (Figure 7). While it would be technically possible to conduct the via Consolare drain water from the Herculaneum Gate, Pompeiian builders constructed the gate on heaped-up earth. Drains are hidden underground below. This downward slope southwards reflects the existence of an underground conduit, which carried away the drain water from the north to the south. The relation between stone-paved streets and subterranean drains was symbiotic, because the drains running under the streets certainly affected the laying out and levelling of the streets in the development of the town (DE CARO, 1992, 23-46). It is notable that not only did the town gate divide the city from its hinterland, but also it blocked the water flow from the outside. There were not any fountains or running water on via Consolare, from the Herculaneum Gate to the intersection with via delle Terme. And the few fountains in this area, despite higher-density settlement, might suggest there were technical difficulties with water supply in spite of public demand. Generally the density of fountains reflects public demand for water; however, this reflection became more significant in the later period. At the initial stage the decisive factor in water supply was technical conditions.

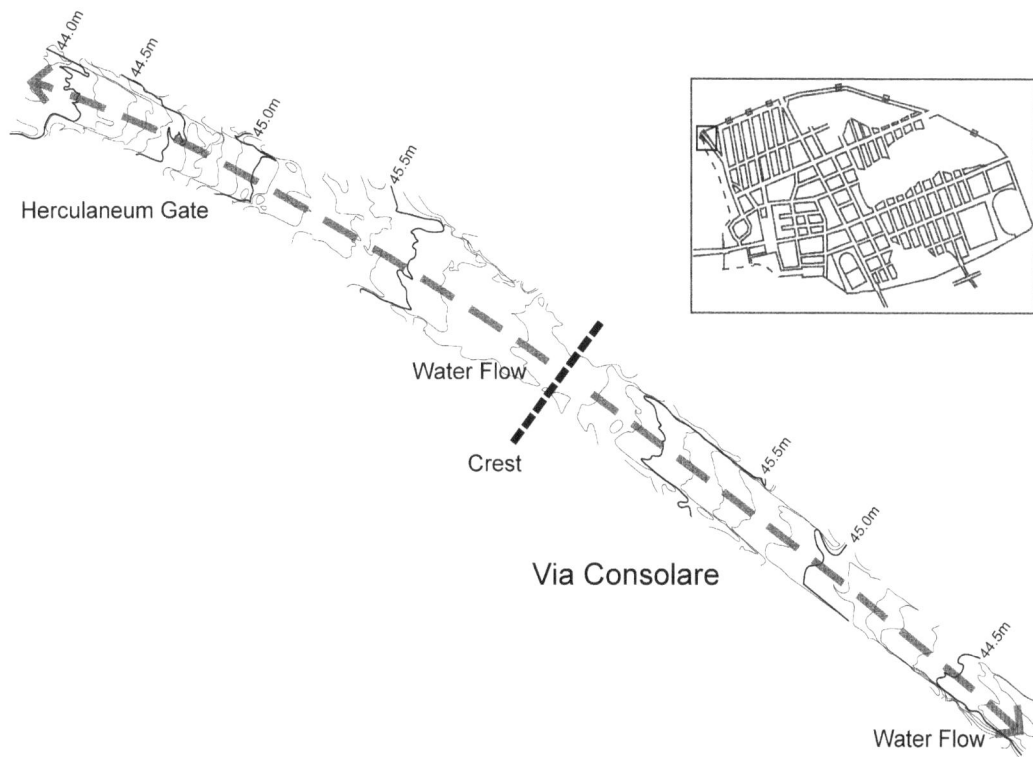

Fig. 7. Crest of the via Consolare and the water flow

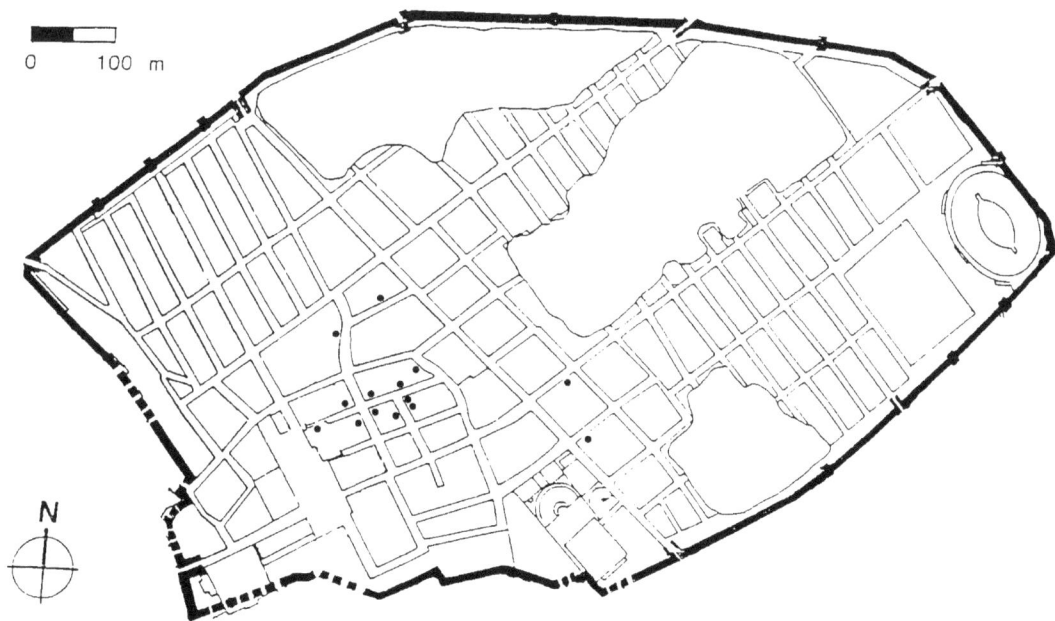

Fig. 8. The distribution of *lanifricariae*
(after LAURENCE, 1994)

Since public water supply was lacking in the area, it is unlikely that excess water from public fountains would wash the waste away from the street. Rainwater in the area to the north of the town wall flowed northwards along via delle Tombe, and the waste from the streets would be collected by hand.

Pipes running from the Castellum Aquae just inside the Vesuvius Gate to all the water towers and street fountains could not provide enough water to the ground area at almost the same level (see Figures 6 and 7). Therefore people living along the northern town wall, with no adequate water supply from public fountains and the run-off after rain, could not expect that water running on the street would flush away their waste from the street.

The network of streets and drainage system in Pompeii

Is there a radical or technical relationship between the network of streets and the drainage system? The surface drainage, as Richardson says (RICHARDSON, 1988, 61), seems to bear little relation to the network of underground sewers, which ran chiefly along the minor streets and alleys and avoided the main axis of the city, perhaps because they were very close to the surface and liable to heavy vehicles crushing their roofs.

In the northern area along the town wall in Pompeii, as we saw above, there was not an adequate water supply from public fountains, although underground sewers were constructed below four out of the six streets running from north to south. Because excreta disposed in cesspits and rainwater on the streets were carried away into sewers, it seems that there might have been little waste on the streets.

Fig. 9. Small elevation with a drain
at the west end of the via degli Augustali

We know the three alignments of streets: via Stabiana; via di Mercurio, supposedly a northward extension from the Forum and the east part of via dell'Abbondanza parallel with via della Fortuna; via di Nola and the vicolo di Mercurio. The west part of via dell'Abbondanza, at the axis (*decumanus*) of the new layout, was the widest street running eastwards from the Forum, and this turned slightly northwards and crossed at right angles by the transverse street (*cardo*), which was the main route for the surface drainage of the town. That part of via dell'Abbondanza that crossed at right angles with the Forum collected all the surface water coming from three public fountains and the streets in the area surrounded by via degli Augustali and the vicolo del Lupanare. Ten *lanifricariae* as well as fullers' dyeworks, the existence of which can be inferred from the strong washing-basins, were concentrated in Regio 7 to the north of via dell'Abbondanza (Figure 8). They need at least brief attention, since they used large amounts of water and discharged waste water onto the streets and, finally, into the sewer. It should be noted that a small elevation constructed on the west side of via Stabiana closed the

entrance to the Forum to wheeled traffic. The original obstacle to wheeled traffic on the Forum was water running on the surface of the street. Pompeiian builders gave a higher priority to effective surface drainage of the main streets than to transport by carriage and carts, even though the former hindered the latter.

The intersection of via degli Augustali (parallel to via Abbondanza) with via Stabiana was enlivened by a water tower, a fountain and an arched portico straddling the pavement. This seems to be a landmark for via Stabiana, characteristic of its destination to the Forum. Actually it had formed an ideal access street for all the vehicles to the Forum, but at the west end of that street the entrance to the Forum to wheeled traffic was closed by a small elevation covering the water-drain conduit (Figure 9). In the final phase, vehicles headed for the Forum turned northwards at the intersection with vicolo Storto or southwards where it crossed vicolo di Eumachia. Surface drainage water, which could obstruct carriages and carts, was provided from a water tower on the west side wall of the Forum Bath and four public fountains. Three fountains had underground sewers, but then could not drain the large amount of water. Vicolo di Terme, vicolo di Soprastanti, via del Foro and via degli Augustali were all rendered impassable to cart traffic in the 1st century BC, perhaps in order to accommodate the crowds that were moving to the Forum. The waste from the streets was flushed away by temporary run-off after rain and storms, as might be expected, rather than by excess water from the public fountains. From the Forum, people could easily continue up to via del Foro, turn westwards on the crossing with via delle Fortuna, and follow the bustling route towards the Herculaneum Gate, or else descend along the attractive commercial zone of via della Fortuna to reach the broad expanses of the eastern areas beyond via Stabiana. The two street fountains are unlikely to have provided adequate flushing water, and via delle Fortuna was designed in much the same way as via degli Augustali.

Town planning and surface drainage in Pompeii

There were many factors that influenced the early development of Pompeii (MAIURI, 1930, 113-286). The first was the quasi-military character shown in the town walls, related to development in the field of engineering. The second was the close connection with the economic hinterland, which increased the population numbers and, consequently, influenced town planning. But looking at the town plan of Pompeii, one is struck by the imbalance between the regular city, stretching eastwards from the Forum in a network of modest urban *insulae*, and the scattered dwellings with well-watered gardens and olive orchards in the west. No doubt the contrast between densely populated urban and scarcely populated rural areas originates from the town planning of Pompeii. Whatever memories of Greek, Etruscan and Samnite old towns may have predisposed the Pompeiian to this regular style, it is evident that the eastern development goes back to a much earlier phase in the town's history. In the early 1st century AD there was some growth of population associated with commercial activity at Pompeii. Still, this settlement stage could not have taken place long after the town planning in the eastern direction. Rather, one must think of urban zones surrounded by a network of streets, which actually functioned as quarters for agricultural purposes. And although one could doubt the likelihood of protecting an agricultural area at a time of population growth, nevertheless agricultural land in blocks surrounded by unpaved streets and bounded by walled enclosures still remained on the eastern side in the final phase. The easiest and most common tactic to drain the standing water was to build a simple level channel down the centre or sides of unpaved streets to carry it away. On the other hand, when owners or occupiers required more space in the eastern *insulae* of via Stabiana in the 1st century AD, they could gain it only by building upwards or by expanding sideways at the expense of their neighbours. It could be argued that the creation of a protected agricultural area in the east preceded the construction of the zone of squared *insulae* along via Stabiana, and that the rectangular agricultural blocks connecting under the unexcavated area offered the necessary interruptions to the squared residential blocks. It thus appears that two distinct types of drains were in use in the two quarters of the city. Each type had a different way of carrying away the standing water.

In the agricultural area, since water is essential not only for the maintenance of life and health in urban areas but also for cultivation, we can expect that the unpaved streets were one of the methods of obtaining water in the intensive agricultural cultivation inside Pompeii. Not only were cisterns important for storing water for living, but also for agricultural purposes in the eastern quarter, as there was not enough drain waters in the streets. Unpaved streets might receive and absorb rainwater (see Figure 3). In the residential areas east and west of via Stabiana and south of via dell'Abbondanza, the water drained from the houses, shops, workshops and gardens, including excess water coming from public fountains, and flowed down the stone-paved streets to carry away the city's waste. In some places on the streets the stones protected the water flow from the traffic. We can therefore look at the boundary of

drainage networks on a crest of via dell'Abbondanza *c.* 100m east from the Holconius crossing between the agricultural and residential areas (Figure 10) to show the way in which the techniques of the drainage system either depend upon or influence the zoning of land use.

The positioning of the new Central Bath under construction in AD 79 was strategically chosen to service as large a part of the urban population as possible. Situated in the block immediately to the north and west of viae Nola and Stabiana, it responded to the needs of the increased population in this quarter of the city, which had led to more waste on the streets. However, the increased population in this quarter resulted not so much in more traffic in the streets round the new bath as in more demand for water supply and drainage. Probably the Central Bath would take over serving the northern and eastern quarters, leaving the Stabian Bath to cater for dwellers in the centre and east. The excess and drain water from the Stabian Bath adjoining via Stabiana led out through the underground conduit rather than joining with the stream on that street.

The impact of the construction of the new bath and probably of the surface drainage network can hardly be underestimated. It becomes evident in the radical and physical changes that took place within the town. Situated in the block immediately to the south of the intersection between viae Stabiana and delle Fortuna, the facilities represented a new town centre project incorporating not only baths but also shops and taverns, and were clearly intended as an amenity for the general population. Those urban changes could have resulted in heavy traffic. In the final phase, the excess and waste water were planned as surface drainage of considerable volume on both sides of via Stabiana. On the other hand, Pompeiian builders avoided the confluence of drain waters coming from the two public baths, which would obstruct the heavy cart traffic on the southern part of the main axis and the south gate of the town. This suggests a major remodelling of the drainage and traffic system in the late 1st century AD.

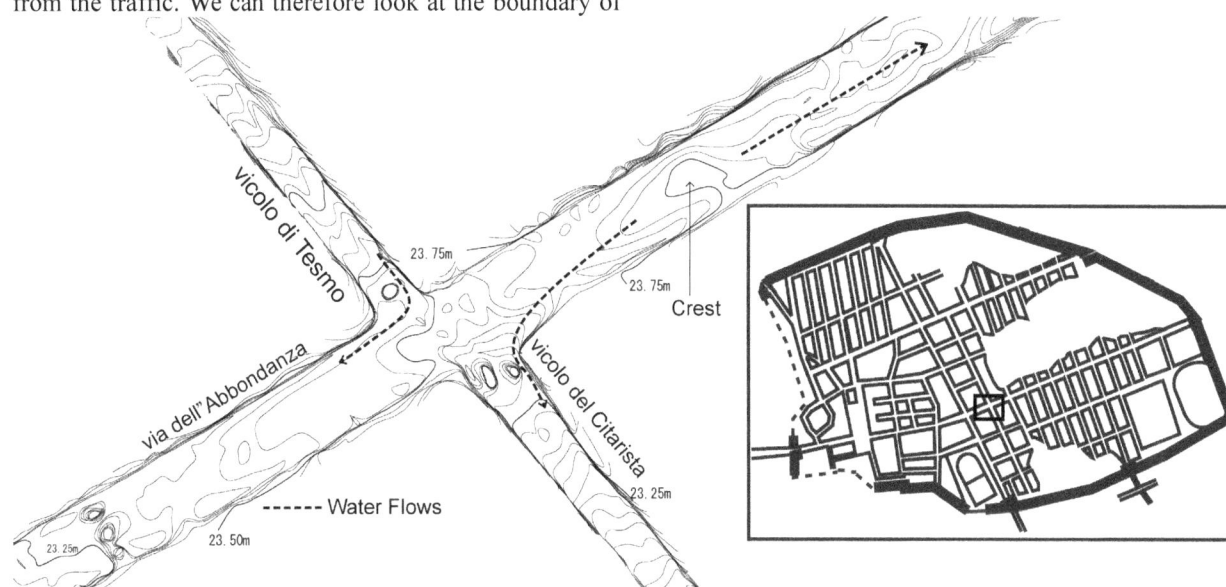

Fig. 10. The water flow on the intersection of the via dell'Abbondanza with vicoli di Tesmo and del Citarista

But such problems could have arisen when the Stabian Baths were first built, because the Forum Bath lacked the capacity to serve the people living in the east. Probably in the beginning of the 1st century AD the Forum Bath was completely repaired to provide increased bathing capacity through a new water supply and drain system. At the same time another large bath complex was built with the underground conduit running north-south parallel with via Stabiana to the east of the developing area; moreover, even when the new Central Bath opened, drainage on the street was not as difficult as it might be expected, because the main underground drain carried into the sewer the waste from practically the entire western half of the town.

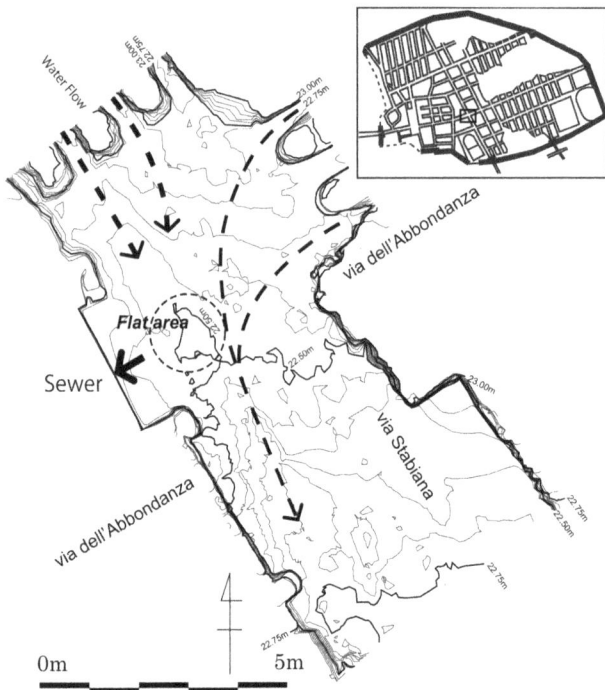

Fig. 11. The water flow on the intersection of via Stabiana with via dell'Abbondanza

The Pompeiians seem to have been particularly successful at water management, because they endeavored to make it functional, strategically located, with water towers, public fountains and sewers. They gave particular attention to areas the Romans neglected in the later period: excess and waste water capable of washing the filth from the streets, which could obstruct carriage and cart traffic. They needed to manipulate the running water to prevent the flooding of some streets and to flush the filth off the streets. One method of managing water was to construct small dikes at intersections between streets to prevent the running water from flowing into them. And another way was to raise and lower the pavement slightly to force the running water to direct its flow as discussed below.

Running water on the streets

The streets in Pompeii were ordered structures, with lava stone for paving its roadway and what may have been pavements and stepping stones on the corners; and a conspicuous feature of the whole site is the elaborate and carefully controlled mechanism of water drainage. The flow of rainwater could cause swelling or settlement of the surface of the streets as water seeks its own level. Here, the water flow on the streets was used by Pompeiian builders, but its use depended on human technology to slightly raise or lower the surface of the streets and thus to control the flow. When public baths and aqueducts were built to provide huge amounts of water, sometimes the small elevation that was raised slightly was constructed by piling an artificial slope against the original configuration of the ground, and the levelling of the surface on the intersections of the routes of the drainage, such as the Holconius crossing (Figure 11), must have been carefully managed to avoid overflow and standing water. The Pompeiians tried to manipulate it to prevent the flooding of some streets and to reduce the flow in others.

A method they applied was to construct a small elevation at the entrance to streets, to prevent the running water from flowing into them. It was, of course, perfectly feasible to discharge rainwater to the outside of the city without directing the water along the streets; indeed there were countless single drainages at Pompeii where water was gathered into underground cisterns by means of simple gutters and drains, but then the water would not wash away the filth without a special method of controlling the flow. At the end of those surface flows, there were really only four gates: the Nola, Sarno, Nocela and Stabiana Gates with surface drains, which could discharge the water running on the streets. Since the gates were closed at night, the running water was led into each drain beside the gate.

Fig. 12. The drain with a revetment outside the Nola Gate

No matter how much those gates and streets may vary in detailed treatment of the structure according to town planning, those four gates share the one similar characteristic of drains beside the gate; with the exception of the Nocela Gate, where they were situated on steep slopes at the edge of the plateau, while at the gates the streets consequently sloped downwards to the outside of the town, the side drain maintaining a gentler and more even gradient than the gateway. At the Nola Gate the drain therefore needed a revetment below (Figure 12).

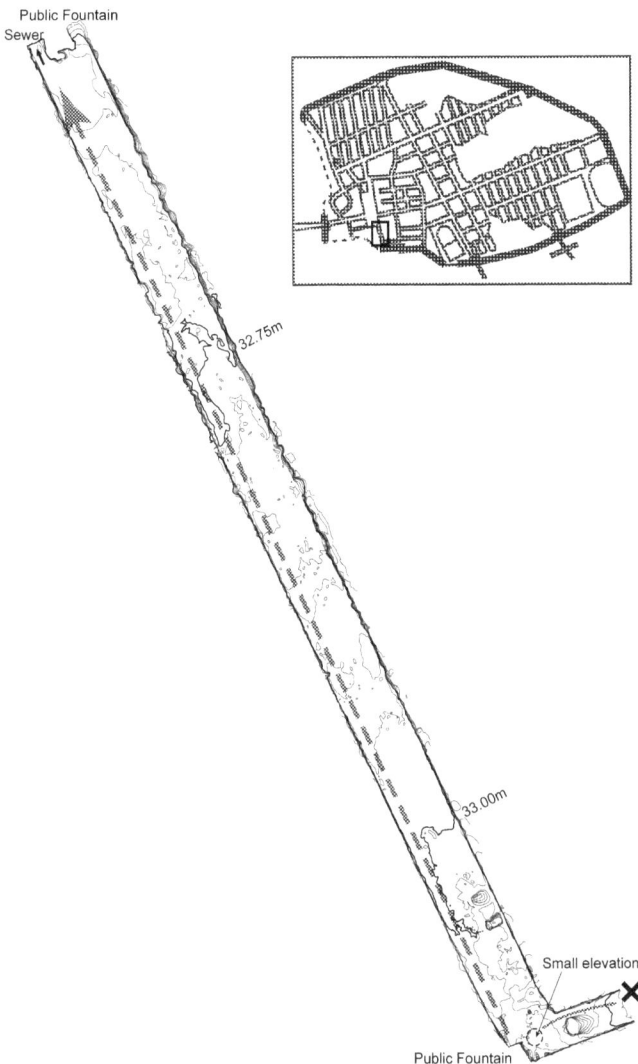

Fig. 13. The water flow on the strada delle Scuole

Pompeiian builders knew that, regardless of where a drain was constructed, the important requirement would have been a constant gradient to keep the water flowing and to ensure that any debris blocking it would not obstruct the drain, into which the surface water on the streets flowed.

Careful construction techniques were necessary for the conduits, drains and sewers, as well as for stone-paved streets; just as on the strada delle Scuole the surface level to the north was approximately 50cm lower than the bottom of the fountain at the south end, thus allowing for the surface water to flow to a drain hole beside the fountain at the north end (Figure 13). There is no obvious reason why the excess water coming from the two public fountains at the north and south ends of the strada delle Scuole evidently were not discharged to the vicolo della Regina, which descends steeply to the triangular square. It is possible that we have here an example of a control mechanism for running water which avoided such a heavy discharge to wash the pavements as would damage the paving stones. It finally ran into via Stabiana, which, along with via di Porta Nocera, functioned as one of the two main drain lines to discharge water to the south; the latter differs from the former in having no adjacent gutter. It would be perfectly ideal and natural to collect rainwater

on the Stabiana Gate at the lowest ground level in Pompeii; indeed there is an adjacent gutter and water channel at the south end of via Stabiana where water, finding its own level, was gathered into this area (Figure 4). However, let us bear in mind that the declivity of the south part of that street from the Holconius crossing considerably exceeds 2.0% (4.7%). The two sewers, which are identified on via dell'Abbondanza and via dell Tempio di Iside west of via Stabiana, could partly catch the rainwater and excess water flowing eastwards from the east half of the Altstadt area through the probable underground conduit running along via Stabiana, so that they have the marked effect of reducing the amount of water discharged through the Stabian Gate.

Conclusion

Basically because rainwater was automatically carried away by the sloping streets, it seems that Pompeii did not need a systematic sewerage system. The run-off after rain that gathered at the Stabian Gate from that quarter of the city could be a problem for loaded carts. The Etruscans solved the problem of sewage with channels covered by slabs in Marzabotto, and the sewers in Pompeii could have been more or less similar to those. Underneath the streets sewers simply covered by slabs were installed. On via Stabiana and via dell'Abbondanza (Sewer holes A and B, see Figure 4), the covered water drain led through the public areas and was maintained as a public sewage system. It was laid out in single lines, rather than as a network, and ran under the streets that followed closely the steeply sloping hillside or valley. Consequently the declivity of the sewer exceeded 4.7% (in parts 8%), which could cause damage to its wall and bed. The sewer underneath via dell'Abbondanza and via Stabiana would have required continued and expensive maintenance, which probably led to the decision to reduce the amount of water running into those sewers. Because at the Holconius crossing a large part of the run-off after rain was carried away by the streets, not by the conduit, there is no drainage hole other than drain holes A, B and C. The sewer underneath the *insulae* and the parallel via Stabiana were used in combination for the drainage system. The builder ensured that secondary drains not feeding the principal drains could likewise handle the extra volume of run-off after storms, following the surface drainage on the highly sloping streets. On the other hand, all the run-off after storms coming from the eastern area of the Forum was conducted into drain hole B linking an underground conduit running under private buildings with the outside of that city. A water-blocking dike built on the north entrance of the vicolo del Lupanare illustrates the control system of running water on the surface of the streets, by means of which the surface water passed from via degli Augustali to via Stabiana and finally to the sewer at the Holconius crossing. The run-off from the northern area that was blocked by that crest on via degli Augustali was not conducted to the drain hole B, but to the drain hole A at the Holconius crossing or flowed along via Stabiana to the Stabian Gate. The drainage network weaves its way though streets and water-blocking dikes. Underground conduits were also constructed, of which the drain holes could collect rainwater. The Holconius crossing (drain hole A) and drain hole B, where run-off and excess water were gathered into the drain hole by means of the streets following the general inclination of the ground, were positioned roughly according to their estimated capacities. According to the capacity of the underground conduit running below the private buildings in Insula 4 of the Regio VIII, the water was gathered through drain hole B from the estimated area along vicolo della Regina, via dei Teatri, vicolo delle Pareti Rosse and via del Tempio d'Iside. On the other hand, following the construction of many private baths and fountains in the northern area of Pompeii in the 1st century AD, the water flow gathering into the Holconius crossing had been steadily rising beyond the estimated drainage capacity of the underground conduit. The Pompeiian builders tried to manipulate the run-off and excess water to prevent the flooding of the southern part of via Stabiana and to reduce the flow through the secondary underground drain, which could obstruct carriage and cart traffic.

The construction of the Central Bath shows how the Pompeiian builders solved the problem of sewage on the streets. On the east side of the Central Bath, three outlet holes can be observed on the lateral face of the kerb stones not wide enough to walk on; vicolo di Tesmo, narrowed by half during the construction of that bath, became just wide enough for one person to walk on but not enough for a cart to roll along (Figure 14). Thus the street became a channel. The water flow, which was blocked by a dike at the entrance to the street along the south side of the insula occupied by the Central Bath, was conducted to via Stabiana through via degli Augustali. It is likely that Pompeiian builders did not intend to level the uneven surface at the crossing of vicolo di Tesmo and via degli Augustali, and the remaining crest of the street was carefully calculated. It brought a constant stream of water into via degli Augustali, which finally joined the main flow of water on via Stabiana (Figure 4). It is not known whether any improvements in drainage infrastructure occurred or any technological development was applied to the drainage system until the building of the Central Bath, when that construction was completed.

The consumption of water must have risen sharply in the 1st century AD, especially with the increase in number of private baths, and would have risen with the construction of the Central Bath. The capacity of the drainage on the streets probably overflowed. Finally, the underground conduits accommodated the different kinds of material that had to be removed: some excess water from the public fountains and run-off after rains flowing on the streets disappeared into drain holes, and some waste water from the baths was carried away into sewers. Pompeii remains an interesting phase of the development of water supply, distribution and disposal arrangements, before the Romans developed and applied systematic and large-scale water technology, mainly in large cities.

Fig. 14. Vicolo di Tesmo (view from the north end)

The streets and sewers in Pompeii appear to be a perfect example of the application of an appropriate technology. This chapter stresses the controlling of discharged water on the roads in connection with the design of its urban structure and shallow gradient, unknown to previous researchers. The criteria for such an evaluation of ancient technology remain unclear; the question arises whether, in the case of regional middle or small class towns, it would not be more appropriate to transfer the updated technological knowledge from the centre of Imperial Rome to that town rather than to improve their own technology.

Bibliography:

BOËTHIUS, A., 1978
Etruscan and Early Roman Architecture, New Haven and London.

CROUCH, D. P., 1993
Water Management in Ancient Greek Cities, New York and Oxford.

DE CARO, S., 1992
Lo sviluppo urbanistica di Pompei, Atti e meomorie della Società Magna Graecia, 23-46.

DOBBINS, J. J., 1997
The Pompeii Forum Project 1994-95, in: S. E. Bon and R. Jones (eds), *Sequence and Space in Pompeii,* Exeter, 73-87.

ESCHEBACH, H., 1979
Die Gebrauchswasserversorgung des antike Pompeji, Antike Welt vol. 10, 3-24.

ESCHEBACH, H., 1995
Die Flächennutzung, in: Eschebach, H. and Eschebach, L., Pompeji, Köln, 103-98.

ESCHEBACH, L., 1996
Wasserwirtschaft in Pompeji, in: de Haan, N. and Jansen, G. (eds), *Cura Aquarum in Campania*, Leiden, 1-12.

ETANI, H. (ed.), 2010.
Pompeii Report of the Excavation at Porta CAPUA 1993-2005, Kyoto.

HOBSON, B., 2009
Latrinae et Foricae: Toilets in the Roman World, London.

JANSEN, G., 2000
Systems for the disposal of waste and excreta in Roman cities. The situation at Pompeii, Herculaneum and Ostia, in: X. Dupré Raventos and J.–A. Remolà (eds), *Sordes Urbis, La eliminación de residuos en la ciudad romana*, Rome, 37-49.

JANSEN, G., 2007
The water system: supply and drainage, in: J. J. Dobbins, P. W. Foss (eds), *The World of Pompeii*, 257-68, New York.

HORI, Y., 2010
Pompeian town walls and *Opus Quadratum*, in: H. Etani, (ed.), *Pompeii Report of the Excavation at Porta CAPUA 1993-2005*, Kyoto, 277-309.

KOGA, M., 1992
The Surface Drainage System of Pompeii, Opuscula Pompeiana II, 57-72.

KOLOSKI-OSTROW, A. O. (ed.), 2001
Water Use and Hydraulics in the Roman City, Boston.

LAURENCE, R., 1994
Roman Pompeii Space and Society, London and New York.

MAIURI, A., 1930
Studi e recherché sulla fortificazione di Pompei, MonAnt, 1930, vol. 33, coll. 113-286.

NAPPO, S. C., 1997
Urban transformation at Pompeii in the late third and early second centuries BC, in: R. Laurence and A. Wallace-Hadrill (eds), *Domestic space in the Roman world*, JRA suppl. 22, Portsmouth, 91-120.

OHLIG, C. P. J., 2001
De Aquis Pompeiorum: Das Castellum Acquae in Pompeji: Herkunft, Zuleitung und Verteilung des Wassers, Nijmegen, 1-33.

RICHARDSON, L. jr, 1988
Pompeii: An Architectural History, Baltimore and London.

SOMMELLA, P., 1995
Urbanistica Pompeiana, in: AA. VV. *Neapolis.*, B. Conticello, Rome, 163-218.

Notes:

[1] For a summary of previous research see RICHARDSON, 1988, 51-63; and for recent discussions with earlier bibliography see JANSEN, 2007, 257-68.
[2] See an important discussion of ESCHEBACH, 1979, 3-24.
[3] MACQUORN RANKINE, W. J. revised by MILLAR, W. J., 1907, *A Manual of Civil Engineering*, 729.
[4] For recent discussion see OHLIG, C. P. J., 2001, 1-33.
[5] The result of that investigation was published in HORI, Y., 2010, 277-309.

THE *CUNICULI* OF TUSCANIA: THE ORACLE DOWN THE WELL

Lorenzo Caponetti
Hypogea - Napoli

Key Words: Tuscania, *cuniculi*, water installation

Underground tunnels dug in the bedrock for water collection or land-reclaiming, the *cuniculi,* are a complex subject to study, one whose boundaries are almost uncertain and that asks the researcher to commit on several levels. Even if only focusing on the hydraulic field alone, we can find that some of them were dug for water capture or as parts of aqueducts, while others were for the generally considered opposite purpose of draining or regulating water level in closed basins; so for a research to be thorough and complete, it must be multidisciplinary. To try to understand the origins of these tunnels means going back to the dawn of agriculture and examining the influence of the irrigation systems on the process of groups forming within settlements and the evolution of society (Liverani, 1998: 3-17); it means trying to identify the moment at which a certain community had the need and the power to put into effect such a modification to the ecosystem. To investigate the way of use of these channels involves analysing the extraordinary heritage of laws, beliefs and customs linked to the finding and sharing of water, and more generally of the allocaion to the individual of the benefits of collective labour. To research how they worked means, ultimately, investigating the reactions of the ecosystem to human intervention, and the ecological meaning of each different hydraulic choice.

The study of *cuniculi* and drainage tunnels is therefore a complex experience: in addition to a systemic approach, it is necessary, more than in other fields of research, to interpose interpretation between the details and the overall vision. There is also the complication of a subject that presents great typological variation and whose complete picture is composed of realities that may appear similar and complementary for understanding although they are chronologically or geographically very distant. Within this scenario, the research that has been conducted in Tuscania has had the benefit of being able to compare the drainage tunnels with the actual instances of oases in different arid countries of the Mediterranean basin. As an extreme environment, the desert oasis represents a great case study for understanding the relation of the cause and effect of human activity on the ecosystem; and the analysis of such a perfect symbiosis between human and environment becomes useful for understanding each intervention's complementarity with the next; these single instances might almost be considered as 'syllables', whose significance cannot be understood in isolation, but through 'language' they become linked to make meaning. Research on *cuniculi* through comparison with similar experiences in other countries has created, maybe for the first time on this scale, the opportunity to widen our vision beyond southern Etruria, extending it to the entire Mediterranean, so as finally to reveal the entire picture: a language with which today we can understand how the Etruscan *cuniculi*, an extraordinary heritage from the past, might represent in fact a great response and an active resource for managing the terrain, and a useful tool for counteracting the effect of future climate change.

Id est foramen

From a structural point of view, the Etruscan *cuniculi* can surely be considered as drainage tunnels: sub-horizontal drainage ways directly excavated under ground, generally furnished with a greater or smaller number of shafts, and whose walls were mainly unplastered when used for transport of water, enabling water exchange along the whole length of the tunnel. Yet the environment and the hydro-geological context in which they operate are totally distinct, not just in the environmental differences between central Italy and the arid countries, but also for the geological complexity of the area in which most of the *cuniculi* are located. The whole area of southern Etruria and archaic Lazio is in fact geologically characterised by a substratum of clay and sand-clay sediments of Pliocenic and Quaternary origin, onto which long and varied volcanic activity has at different times deposited a compacted layer of both explosive and effusive materials, thus creating a complex situation that is made up of many heterogeneous horizons in which permeable substrata are continuously alternated with palaeo-soils and other impermeable formations. This complexity of geology is evidently reflected in the underground water available: volcanic materials, due to their strong dishomogeneous qualities, are generally characterised by water circulation over many near or distantly communicating horizons, while the levels of low permeability, even within the volcanic formations, are responsible for a 'suspended circulation' which has a locally varied area extension, and a strong lateral discontinuity. The point of contact between volcanic and impermeable material, instead, can be considered the bed of the 'basal stratum' which, compared to the suspended stratum, presents a higher potentiality and a greater area of distribution, and is therefore the main water table exploited by the wells.

Fig.1 General view of a *cuniculus*
(The length of this kind of tunnel in Etruria is in most cases between a few tens and few hundreds of metres, making them a lot shorter than most constructions in arid countries. Some are known to span some 3 miles or more, however.)

Within this complicated hydro-geological situation, it is understandable how what Pompeius Festus described as a *foramen*, nothing more than an underground cavity, can be used in many different ways to enable or increase the ease of hydraulic communication between various areas, and to allow or facilitate water exchange and circulation. The many and various needs felt each time to regulate or to drain higher areas, as well as to supply lower areas, have determined a range of applications and typologies so vast and varied that, despite many attempts to classify the various typologies (DEL PELO PARDI, 1943, JUDSON, KAHANE, 1963, 89, RAVELLI, HOWARTH, 1988, 58),

we often find it hard to classify each case study with certainty within one particular class, especially in those cases in which opposing views do not allow a definite determination of the entire picture.

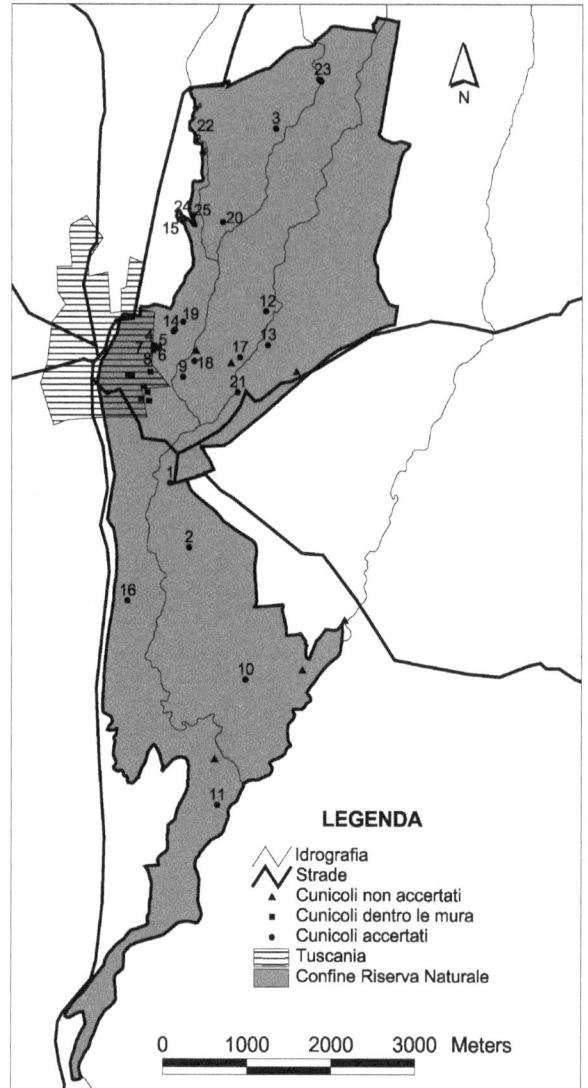

Fig. 2 Map of the *cuniculi* in the Natural Preserve of Tuscania (after CAPONETTI 2006, 20)

(Beside the contours of the present-day urban centre, the map shows main roads and streams. The *cuniculi* are divided into ascertained (round dot), reported but not ascertained (triangle), and those within the former city walls (square).

In addition to the *cuniculi* excavated for water capture - whether as a means of tapping the water table or collecting rainwater - many were found on excavation to fall within the wider scope of 'drainage', ranging from reclaiming marshland or small closed basins, to the lowering of saturated areas of water-drenched land, following a principle similar to the one adopted nowadays of the so-called mole-plough (QUILICI-GIGLI, 1983, 116). Speaking of drainage works, it is worth remembering those of Veio, excavated for miles along the valley's axis, to substitute for surface streams to prevent or limit superficial erosion, and enabling a radical transformation of agricultural land (JUDSON, KAHANE,

1963, 90). Sometimes they are used to modify or implement surface streams, regulating lakes or connecting crosswise parallel valleys, to dry out one or to bring water to the other. In other cases the same principle is found to be used, on a smaller scale, to divert part of a surface stream underground, making a by-pass which enables building over the stream, instead of a bridge. This is a solution adopted in many cases during the construction of roads - like for example the well-known Ponte Terra near Tivoli (CASTELLANI, 1999, 109-28); or even the present road between Tuscania and Viterbo, which after 9 kilometres still crosses the Lemme stream thanks to a water tunnel of this type; but also the possibly sole pre-Roman aqueduct (WARD-PERKINS, 1962, 1642-3) of Ponte del Ponte near Corchiano, in which a massive wall, which in this case does not sustain the passage of a road but the crossing of an aqueduct, uses arches instead, which the Roman aqueducts have accustomed us to. Moreover, many are the cases in which water tunnels are used to transport and distribute water either diverted from streams or for servicing houses in an urban context - like the well-documented case in Orvieto, for example (BIZZARRI, 1991, 165 and picture 120); or finally for the simple function of storing water, a purpose achieved either by excavating tunnels radiating from the bottom of wells, or by partly closing the mouth of a tunnel with a wall, to hold water up to a certain level.

Sacred uses

Cases are known, in the end, where *cuniculi* were dug as part or whole of various kinds of places of worship, namely as the water producing part of an above ground sacred site - see for instance the one at Veio (RAVELLI and HOWARTH, 1988, 66), or the temple of Giunone Curite (FELICI et al., 1994, 27-32) - or as parts of funeral complexes like the so-called Labirinto di Porsenna at Poggio Gaiella in Chiusi (RASTRELLI, 2000, 96-105), when not as underground sacred sites *per se*, as appears to be the case, for example, with the so-called Tomba della Regina in Tuscania, with its three-layered labyrinth of *cuniculi* (BERNARDONI and ORLANDI, 2002). It is worthy of notice, by the way, how many of these devices, both of the first kind or the second, at various times during history were the origin of places of worship for communities completely disconnected from the ones who originally excavated them, to the point that even today many of them are related to temples still in use today. Speaking of this, it is worth mentioning the debate that came up even in recent years on the supposed influence of the *cuniculi* on the origins of catacombs. The question was brought up in earnest in the 1970s by Francesco Tolotti, a scholar who perceived a former hydraulic function and a subsequent distance between the original excavation and the following funereal use in a number of cavities, if not entire regions of catacombs in Rome, among which, and only to name the most important ones, are the central region of the Pretestato cemetery (TOLOTTI, 1978, 161-4 and 171-80) and the principal gallery of the Nicomede hypogeum (TOLOTTI, 1978,

164), and, not long after, both the initial region, the Acilii hypogeum and the second level of the Priscilla catacomb, the cavity of the Good Shepherd and the Flavii hypogeum at Domitilla, the initial cavity of Calepodio and the whole catacomb of S. Tecla (TOLOTTI, 1980, 25-32 on Priscilla; 34-8 and 41-2 on Domitilla; 36 on Calepodio and 37-9 on S. Tecla. Also in the same paper more on Pretestato at pp. 32-4).

Although Tolotti was not the first to see a former hydraulic function in parts of the cemeteries of Rome, he was the first to speak openly about hydraulic hypogea as possible precursors of the cemeteries. In his opinion, the influence of hydraulic works on the origin of catacombs was more than just a work hypothesis: not just, he stated, 'a few of the underground cemeteries of Rome - among which more than one of the early ones - originated from abandoned hydraulic works' (TOLOTTI, 1980, 43), but even 'the first cemeteries that were excavated as such must have been conceived and carried out with the essential help of technicians and workers with a background in hydraulic works' (TOLOTTI, 1980, 45). Not just a continuity in the use of some of those hypogea, converted at some point from water storage units into a sacral place, but also a connection through the people who were in charge of hydraulic works, whose technical skills, according to him, were crucial for the excavators of the first cemeteries, whether they belonged to the same community, or were hired as an external source of skilled labour. The work of Tolotti was highly criticised, and a great part of the more recent literature raises doubts about most of his analysis of all the various hypogea he used to support his ideas. In the words of Philippe Pergola, 'though fascinating and sometimes archaeologically documented, [Tolotti's theory] is in most part a pure work of hypothesis, not supported by solid evidence. Sometimes the early hypogea are located at such depths as to reasonably exclude that they could have originated from nets of tunnels that were definitely running several metres above, where large nets of still accessible *cuniculi* have been found, completely ignored for hundreds of metres by the funereal tunnels, that only intersect them by chance: it is rightly the case, for example, with one of the cores of the catacomb of Domitilla mentioned by Tolotti' (PERGOLA, 1997, 64). Even more recently, a criticism of Tolotti's reading of many of the structures in Pretestato came from the thorough, punctual and well documented work which Lucrezia Spera did on that complex (SPERA, 2004, 11-20), after which Tolotti's hydraulic model for that cemetery must be drastically downsized. The same, according to her, should apply to most of the structures that Tolotti analysed in the Nicomede, the Priscilla, the Domitilla, the Calepodio and the S. Tecla cemeteries (SPERA, 2004, footnote 88 at p. 15). U. M. Fasola and P. Testini, on the same subject, though not rejecting the core of Tolotti's thinking, pointed out his tendency to identify as hydraulic even structures that certainly were not, ending up by drawing a picture far too complex and contrived to find comparison in the panorama of Roman hydraulic works (FASOLA, TESTINI, 1978, 124, quoted in SPERA, 2004, 15 and footnote 88 on the same page).

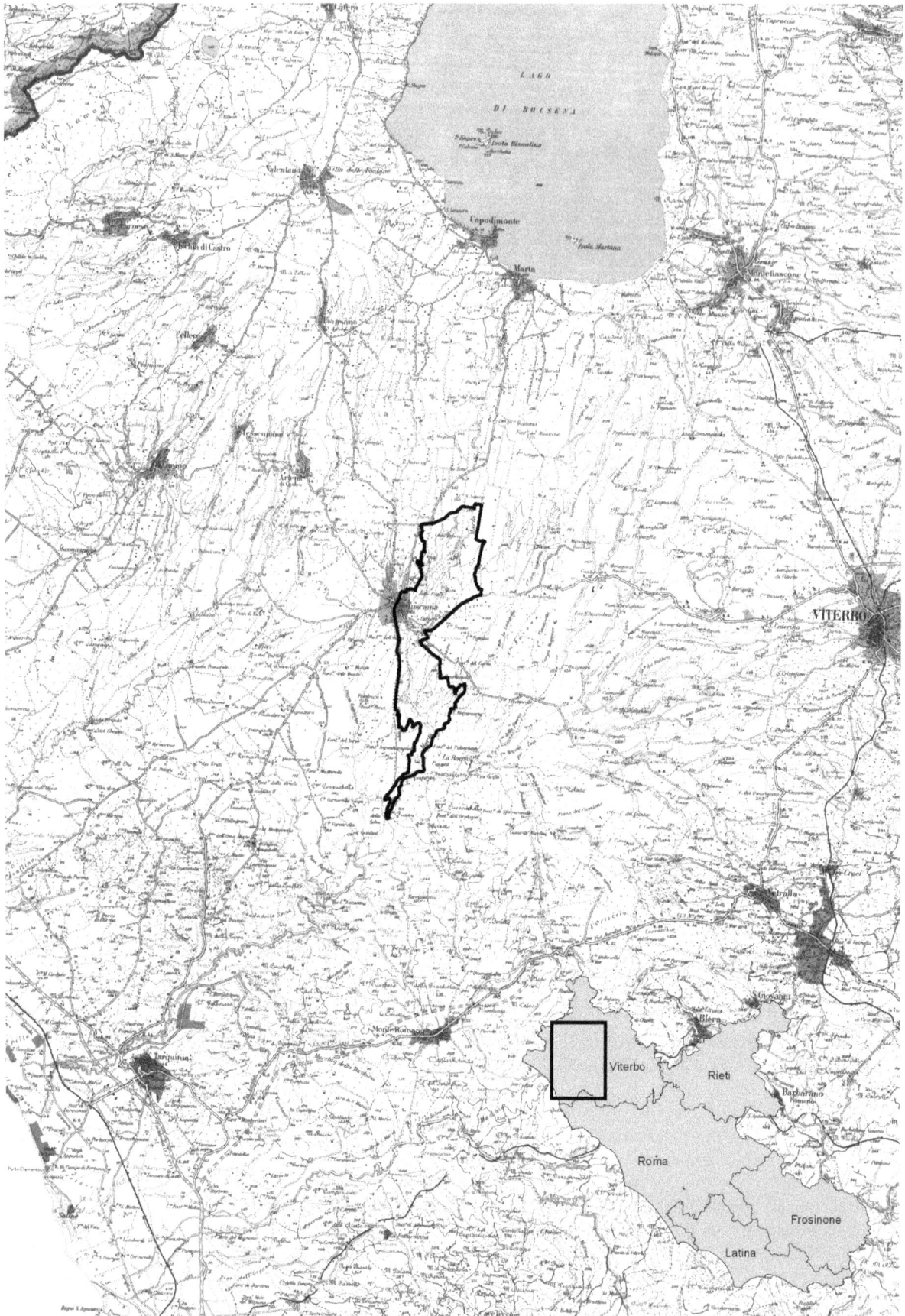

Fig.3. Regional map of the former southern Etruria, with an indication of the surveyed area
(after CAPONETTI, 2006, 18)
(Tuscania is about 50 miles north of Rome, right in the centre of former southern Etruria, at the edge between the volcanic area of the
Monti Volsinii and the low-lying Pliocene clays.)

		Cunicoli found within the boundaries of the Natural Preserve of Tuscania		
N°		Name	Vicinity	Supposed main function
1		**San Potente**	Necropoli di Sasso Pinzuto	Water collection
2		**Quarticciolo**	Quarticciolo	water collection
3		**Acquaforte**	Acquaforte	water collection
4		**Bottacce 1**	Bottacce	water collection
5		**Bottacce 2**	Bottacce	Water storage + ?
6		**Bottacce 3**	Bottacce	Water collection + distribution
7		**Bottacce 4**	Bottacce	Water collection (?)
8		**Bottacce 5**	Bottacce	water collection
9		**Maschiolo sx 1**	Valle del Maschiolo	water collection (?)
10		**Forma Italiae1**	Quarticciolo	?
11		**Cunicchio 1**	Cunicchio	water collection
12		**Pian di Mayere 3**	Pian di Mayere	?
13		**Forma Italiae8**	Curva dei pali di Ferro	?
14	a	**Forma Italiae9**	Cimitero	Transport (?)
	b	**Forma Italiae9/b**	Cimitero	transport (?)
15		**Forma Italiae10**	Castelluzza	transport (?)
16		**Ara del Tufo**	Ara del Tufo	water collection
17		**Pian di Mayere**	Pian di Mayere	Water divertion (?)
18		**Maschiolo sx 2**	valle del Maschiolo	water collection (?)
19		**Peschiera**	Peschiera	water collection
20		**Castello Broco**	castel broco	water collection
21		**Scalette**	Necropoli delle Scalette	water collection (?)
22	a	**Maschiolo dx 1**	Pantalla	water divertion
	b	**Maschiolo dx 2**	Pantalla	Transport + water collection
	c	**Maschiolo dx 3**	Pantalla	transport + water collection
23	a	**Acquarella 1a**	Salumbrona	water divertion
	b	**Acquarella 1b**	Salumbrona	water divertion
	c	**Acquarella 2**	Salumbrona	water divertion
24		**castelluzza 1**	castelluzza	?
25		**castelluzza 3/Maschiolo**	Castelluzza	?
		Rivellino	Within the city walls	transport (?)
		strada S.Maria snc	Within the city walls	transport (?)
		Don Pino	Within the city walls	transport (?)
		Lavatoio	Within the city walls	transport (?)
		Sette cannelle	Within the city walls	transport (?)
		Santa Maria	Within the city walls	water collection
		Strada S.Maria 18	Within the city walls	transport (?)
		Cunicchio 2	Cunicchio	Unknown
		Forma Italiae6	Strada per viterbo, km 19	?
		Pian di Mayere 2	Pian di Mayere	?
		altro lungo Maschiolo	valle del Maschiolo	water divertion (?)
		capanna di sasso	Poggio Colone	?

Fig.4 Chart of cuniculi within the perimeter of the NPT. In bold the ones ascertained.
(after CAPONETTI, 2006, 21)

Fig. 5. Inner view of a highly eroded *cuniculus*
(The openings of the shafts, at exactly 35.50 m apart – one roman *actus* – are in this case a valuable hint for dating the work. Clues are generally quite rare in works that were purely functional and that share excavation techniques common to almost all the pre-industrial era.)

Seepage

Alongside Tolotti's enthusiasm for hydraulic works and subsequent over-sampling and misinterpretation, one has the feeling that he clashed on one side with the general reluctance of mainstream thinking of the time to consider pagan origins of many catacombs, and on the other with a tendency to underestimate the typological variety of underground hydraulic works, to give little importance to thoroughly analysing from time to time the possible functioning of each of them, and to designate them all as cisterns for water storage or *cuniculi* of 'agricultural use' (PERGOLA, 1997, 64).

In most cases, non-Christian or non-Jewish early hypogea have been considered part of later Christian or Jewish catacombs, and it has turned out to be difficult in most studies of the past twenty years to relate the initial core of the different cemeteries to one particular community or religious group (see PERGOLA, 1997, 57-8 and 61, and more recently PERGOLA, 2006, 177-82). The misconception in classifying former hydraulic works, and the general assumption that their use or function must in any case have come to an end before they were converted to funereal use, may well be the case for most if not all of

the above-mentioned works that were in fact cisterns, and they had run out of use long before the excavation of the first grave. There are other cases, though, in which a comparison with other findings both within the field of works that were converted into cemeteries and in the case of drainage tunnels may suggest that there are reasons to think this is not always the case. Beside Rome, even in southern Etruria, an area geographically, culturally and historically close to the capital city, there is 'ample documentation of the re-use of hydraulic works [...] in the excavation of catacombs. [...] The use of pre-existing hydraulic works in our territory is even documented [beside the catacombs Ad Bivium on the Via Cassia and Ad Vicesimum on the Via Flaminia] in the hypogea of Sorrina Nova, of Santa Cristina at Bolsena, of Nepi, of the 13,500 km of the Via Tiberina and of Monte della Casetta' (FIOCCHI NICOLAI, 1988, footnote 1636 at pp. 369-71. The same text, at pp. 128-9, 149-53, 249, 334 and 343-4, reports descriptions of all of the quoted hypogea.) In the one at Santa Cristina, from what has been surveyed of it, the plan of the hydraulic works shows a main single gallery with possible branching out at the two ends that looks to be a lot closer to a captation *cuniculus* than a cistern. This may turn out to be purely conjectural, though, given the modifications it went through and the fact that it is not fully accessible today. (On the Santa Cristina catacomb, see FIOCCHI NICOLAI, 1988, 149-53; see also map at page 138, fig. 95.) Besides, these kinds of works are known to be poly-functional, and to speak about the one purpose for which a particular cavity was dug can sometimes be misleading. Moreover, any modification of a subterranean structure, for its own nature, is very likely to destroy or obliterate it to the point of making it impossible to read any earlier phases (PERGOLA, 1997, 58). What is strikingly different from any cisterns, however, is the so-called hypogeum of Riello (also known as Sorrina Nova), a small underground structure on the outskirts of Viterbo, in which a former labyrinth of *cuniculi* only partly underwent conversion into a catacomb (FIOCCHI NICOLAI, 1988, 127-30), and in which the former layout of frequently meandering tunnels, the walls unlined, demonstrates a lot more connections with the Tomba della Regina in Tuscania than with any cistern around central Italy. Besides, in both cases, there is water even today, something that may lead us to assume it has been there continuously since the excavation of the cavities, and hence even when these were converted into a cemetery. Lastly, as is so often the case with all archaeological research, the lack of data may have played a role in this supposed underestimating of hydraulic works. It is the case, one more time, with the *cuniculi* net which Tolotti mentioned as pre-existing and obliterated by the expansion of the Priscilla complex: recent findings seem to prove a flourishing phase of agricultural plants in the same area, whose chronology may provide hints to comprehend a work that 'has been so far too quickly dismissed by some authors as simply a work of hypothesis or as an exuberant pan-hydraulicism' (GIULIANI, 2006, 166 and footnotes 10 and 11, on same page).

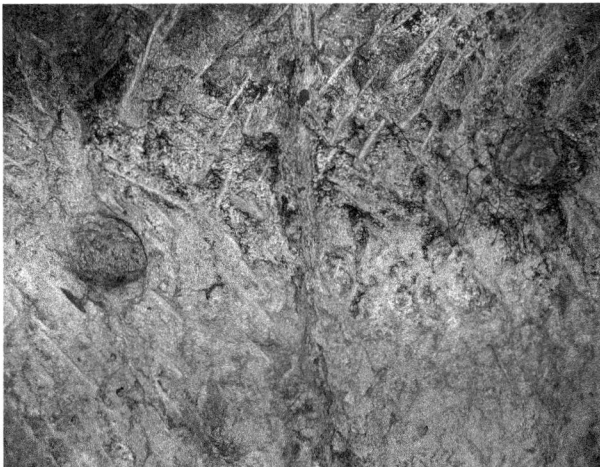

Fig. 6. Detail of a *cuniculus* wall
(Beside two lamp niches, pick marks are clearly visible in the top part of the picture. The marks coming from both the left and the right side show that this was the meeting point of the two excavation teams. Starting the excavation simultaneously at the bottom of two shafts is a common technique in most countries, in some cases still in use today.)

All of this, far from proving any of the authors wrong who have been quoted so far, may inspire a few thoughts and provoke more questions on the subject: it seems that the possible influence of hydraulic works on the origin of catacombs would need to be researched more, and it seems that the knowledge of drainage tunnels could prove useful for this. In all those cases in which the understanding of former hydraulic structures is not fully achieved, knowledge of how drainage tunnels' functioning relies on seepage, condensation and the integration of many small, undetectable movements of water may contribute to casting new light on at least some of these cases. Moreover, if any of the hydraulic works which initiated the catacombs prove to be or to have been reliant upon the same kind of 'hidden rains', can we imagine that that place might have been considered sacred by somebody at some point in history, like many similar hypogea were and are across the Mediterranean? (LAUREANO, 1995, 215-52) Lastly, if any of this proves right, are we allowed to see, in the converting of these into cemeteries, a continuity in the captation use and subsequent continuity of worship by different communities and religions over time, instead of a discontinuity with water storing devices that supposedly stopped being used? There is very little archaeological evidence, if any, to support all this right now. And yet there are hints that seem to be pointing in the direction that this may be plausible, and that the evidence is absent only because nobody has searched for it yet. As far as I am concerned, this should be enough to keep us searching.

San Potente

The small *cuniculus* of San Potente, located about 1 kilometre south of the town of Tuscania, along the track of the ancient Via Clodia, with its water constantly flowing even today during all seasons, can be considered an example of the many water tunnels spread around the region, and of how they are still useful to people nowadays. But if it is easy to associate their continuous functioning with the maintenance and adjustments done by those who felt the need of water over the course of time, it is more difficult to understand the needs for which they have been excavated, which community supported the work, and in which age this happened. Although each *cuniculus* may give 'many more clues than one could investigate over many years of study' (CASTELLANI, 1999, 56), it is also true that in most cases these are just functional structures, lacking any kind of stylistic characterisation, with construction techniques that are basically constant during all of the pre-industrial age, and that in many cases have been subjected to continuous use, which therefore disguises their period. Our understanding cannot come from the analysis of a single model but from a diachronic and regional analysis of the surrounding territory. This research may, nevertheless, clash with the lack of literature on landscape archaeology and the development of normal agricultural practice: from this point of view the main problem is not a shortage of information on the tunnels themselves but the lack of studies of 'ordinary and everyday life, the archeology of "the people without a story", rather than "the people who made the story", who have, due to tradition, dominated most archaeological and historical surveys' (BARKER, 1987, 29). This condition is strongly applicable to Tuscania, made up of small and independent examples, structures easily created in a short time by a rather small community, if not the result of a single family's effort, and which therefore cannot be attributed to a particular historic phase, as has been done at Veio (JUDSON and KAHANE, 1963), or in the territory between Cisterna and Velletri (QUILICI GIGLI, 1983), where the extensive *cuniculi* nets, organised as part of an important work of land reclamation, have been related to the presence of a central power able to plan and carry out a work of such importance.

Fig. 7. View out from the *cuniculus*
(The section is seldom wider than was strictly necessary for the work: 5 to 6 feet high, 1.5 foot wide, and barrel-vaulted top are characteristic features of by far the greatest number of *cuniculi*.)

Placed somewhere between 'the presence of stable settlements with a population engaged in consolidated agricultural activity' (RAVELLI and HOWARTH, 1988, 63) and the realisation of complex hydraulic works of a well-known age, the origin of *cuniculi* seems to be linked to the important work of land reclamation that enabled the Etruscans to cultivate and master huge portions of territory, one of those elements which were 'at the same time cause and effect of structural changes, assuming and demanding different types of social cooperation' (TORELLI, 1981, 23). At the moment, there are no elements to indicate the presence of foreign workers and their technical know-how, like the ones for example known at Tarquinia since the first quarter of the 7th century BC and responsible for the technique of 'pillar walls' (BAGNASCO GIANNI, 1996, 39), or the Eubean

artisans who seemed to settle in Veio and Tarquinia after the first half of the 8th century BC, and thanks to whom, presumably, not long after, refined Greek ceramics were found around south Etruria (D'ERCOLE et al., 2002a, 131). On the other hand they appear to be a local invention instead, an intuition linked to the tufa capacity of absorbing each year nearly a third of the rainfall (Boni et al., 1988), and then releasing it slowly, to the foot of the many vertical cliffs, at the contact point with the Pliocenic clays underneath. Might they, or not, be linked to Eastern know-how, and therefore with the origin of *qanat* and other drainage tunnels; the *cuniculi* genesis has to be taken into account with their rather fast evolution since the formation of the big proto-urban centres of the 11th and 10th centuries BC (D'ERCOLE et al., 2002b, 111), and between the 9th and 7th centuries BC with the surfacing of the Etruscan state system and the systematic propagation of arboreal cultivations such as the vine and olive tree (BARKER, 1987, 28), a clear sign of territorial conquest. At this moment, again, we do not know at which point of this important period of immense changes the *cuniculi* technique actually began; it is hard to tell exactly which of the changes that in Etruria led to the development of urban centres, to the introduction of writing, of patronymics and private property, have to be preceded or followed by the intuition that by excavating a particular tunnel under the ground, water could be exploited. What we can by now be reasonably sure about is how strongly Etruscan hydraulic engineering influenced the urban organisation of archaic Rome, and the substantial continuity that links it to the first realisations of Roman hydraulic engineering: from the first layout of the area that will host the Forum, to the realisation of the Cloaca Maxima, and to the excavation of the Appio aqueduct - the first of the big such ones in Rome, which for over 16 kilometres runs along an underground tunnel: it is nothing but the application of similar solutions, adopted more and more on a wider scale, along a route that, following the systematic adoption of the arch as a mean of technical construction, will lead to the master works of Roman hydraulic engineering.

Zeravshan

About 200 km north-west of Samarkand, in present-day Uzbekistan, the oasis of Nurata is nowadays a town of some 25,000 inhabitants with a soviet appearance, and where little remains of the charm of a town said to have been built by Alexander the Great. Nevertheless, it is enough to get out of the city to encounter the long and numerous series of craters left by the collapse of the *karez*, the drainage tunnels which until recently provided water for irrigation and other human activities, and which, abandoned in the incoherent alluvial geology, have not resisted the action of time. In fact, independently from their antique origins, that a local tradition as uncertain as it is plausible wants to date back to Alexander, the *karez'* decay is recent. Starting from the agricultural reform demanded by the soviets in the 1960s, who to expand the cotton monoculture destroyed the region's agricultural landscape totally, it has still not completely ended today: the so-called *Kalta-karez*, 4 kilometres from the city, known to be the last channel still functioning, still has a certain quantity of water; but in the last few years the family that looked after it has moved away, so in certain points the gallery has started collapsing, and the interrupted water flow has lost its transporting capacity, beginning to deposit detritus within the gallery, instead of keeping it clean by carrying it outside. The lack of use and maintenance means that the survival and functioning of the tunnels are threatened, also because their original purpose is so easy to forget, once the system does not produce enough water to be used, and is therefore considered useless.

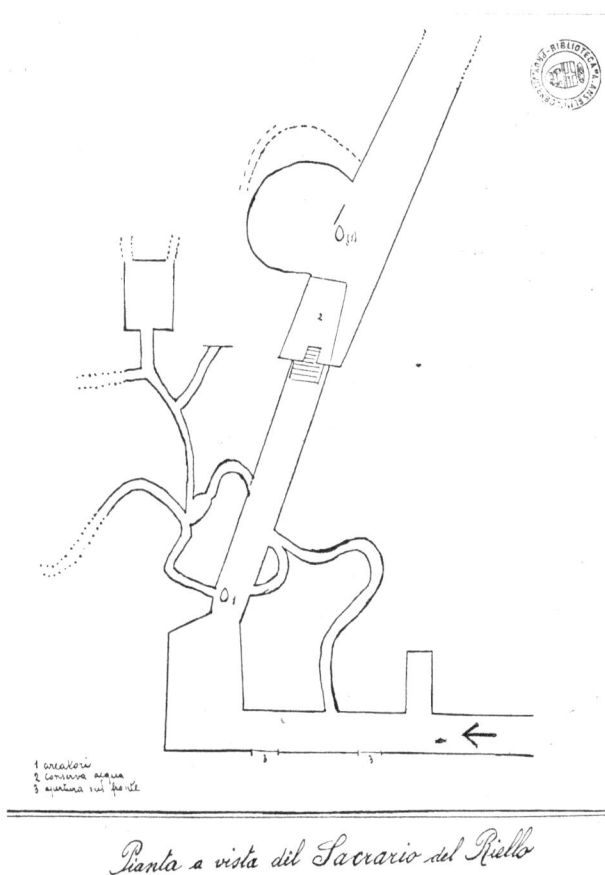

Pianta a vista del Sacrario del Riello

Fig. 8. Map of the Riello hypogeum (after SIGNORELLI, 1966, 141)
(Although the accuracy of the details is probably questionable, the map by Signorelli is still quite effective in showing the meandering of the labyrinth and the squareness of the works that cut through it to convert it into a catacomb.)

A relationship which is clear instead in all those cases in which, thanks to the correct adoption and functioning of specific solutions, human intervention has enabled the creation of an agricultural landscape capable of production even in the harshest conditions. In those cases in which human action integrates itself correctly with the natural environment, the result is as clear as the turning of a desert into a garden, and the importance of this relationship it is not only unmentioned, but it is laden with other meanings too: this is the case in Samarkand, where quite deliberately the river that guarantees life and

sustains all the oasis' production has been given the name of Zeravshan, 'the gold bearer' in Persian. But it is probably the case for Tuscania as well, where in the resort of Bottacce, at the entrance to the town on the side of the Maschiolo valley, within just a small portion of land there are five *cuniculi*, apparently independent from each other, which give water to an extraordinary terraced vegetable garden complex, and whose historic presence, over here too, reflects in the name of the area: apart from the actual name, which nevertheless referred to the hydraulics and to nearby Moletta, placed significantly further down to use its waters, all the area was known in the past as the Valle Aurana or Valle dell'Oro, a name which probably referred not to a hypothetical 'Aureo, King of Etruria', as suggested by the 16th century Tuscanian scholar Francesco Giannotti, but more likely to the fact that, to use his own words, in it 'there are very many other springs, with very ancient and beautiful subterranean aqueducts and caves' (GIANNOTTI, 1606, 24). The Valle dell'Oro's origin and location have been obscured by collective memory, despite it still actually functioning, and despite it living on through each street corner of the district that was named after it; this is a clear indication of how, more than in the case of principal monuments in the town, the relationship between the *cuniculi* and the cultural landscape has to be researched through the extraordinary continuity of use of most of its systems. This is impressive testimony to how the population has freely recognised the effectiveness of its adopted solutions, and has freely volunteered in almost half of the cases for the maintenance works necessary to perpetuate its functioning. And if the Valle dell'Oro of Tuscania is still a live reminder of the past mutual relationship between human and environment, a public acknowledgment of how rich and generous the natural environment can be if modified to suit us, the *Kalta-karez* of Nurata is a clear example of how the two symbiotic partners are in fact tied together and interdependent; of how it is necessary, for that particular agricultural landscape to keep existing and sustaining humans, that humans look after it on a daily and constant basis.

Tam opertus quam apertus

The research conducted around the town of Tuscania has demonstrated how the *cuniculi* are in fact a great work of water regulation, an infrastructure capable of enhancing the relationship between underground channels and the water cycle, enabling sustainable management of water resources. This is a network whose functioning is based on the complementarity of many small interventions, which clearly demonstrates the continuity between above- and under-ground operation of the water cycle, and which - like the oasis - is a clear result of a human-environment symbiosis. Even more than the technical difficulties encountered by the construction of any single work - for example the channel out of Lake Albano, a straight tunnel, almost 1.5 kilometres long, excavated probably during the archaic age below lake level (CASTELLANI, 1999, 57-74) - what strikes the imagination is the skill of intervening at the interface between surface and underground water, a sign of a

21

perfect knowledge of how these are the two sides of the same water cycle, and a clear indication, on the other hand, of how distant are the solutions adopted nowadays, when 'in many sectors of hydraulic engineering, projects and constructions are created as if the surface and underground water systems were completely independent' (CUSTODIO, LLAMAS, 2005, 258). This is an infrastructure based on the exclusive use of natural dynamics, operated by nothing but gravity, and which in the systems' linearity, in the high and porous tunnel walls, finds a perfect strategy to enhance the exchange surface between above ground and underground. This functioning by gravity, besides not requiring any kind of energy to bring water to the surface, implies that the water used is the sole supply the system is able to give, preventing over-exploitation and therefore guaranteeing sustainability. The linearity of tunnels and canals, as opposed for example to the punctual captation of a vertical well, guarantees the possibility of exchange, which - moved by passive forces - can reach a bigger volume, necessarily needing longer time or wider surfaces. What better proof of how sustainable linear and gravitational captation is than the fact that the small *cuniculus* of San Potente is still giving water centuries after its excavation? Just like this example, in most cases the *cuniculi* of Tuscania are still functioning, or could easily revert to their original use, should the initial conditions be re-established. The vertical wells, provided with machinery operated on the basis of choices that are purely economic and therefore short-term related, have the danger of over-exploiting the water resource, causing the lowering of the water table, with the risk, within a short time, of having to excavate the wells ever more deeply. And it is not just a matter limited to issues of energy or volume of demand.

The network of tunnels dug just a few metres under the ground in fact rarely intercepts the main water table, and whether due to rainfall, or condensed water, or other kinds of hidden rains or trickles between the cracks, most of the time they rely on surface aquifers or recharge areas, generally local and of modest size. This implies a direct relationship with a small area of land immediately above or nearby, which in turn eliminates any problems caused by sharing a common, large and incontrollable resource such as a water table, which is a potential cause of tension, even at international level. Besides, it is worth noting how gravity systems are usually an expression of a community, rather than an individual, and be it the descendants of those who excavated the tunnel or the members of a union, the collective interest will generate shared laws that will enable survival of the resource in the future, while the individual interest would be to exploit it as much as possible in a very short time. But the fact of utilising local water also means dealing with a relatively small ecosystem, with limited inertia and resilience. That is why any change made to the system, even on a small scale, can potentially imbalance and destroy the tunnel ... and back we go to the concept of the oasis, and of how each element has a well-determined role in maintaining the general balance.

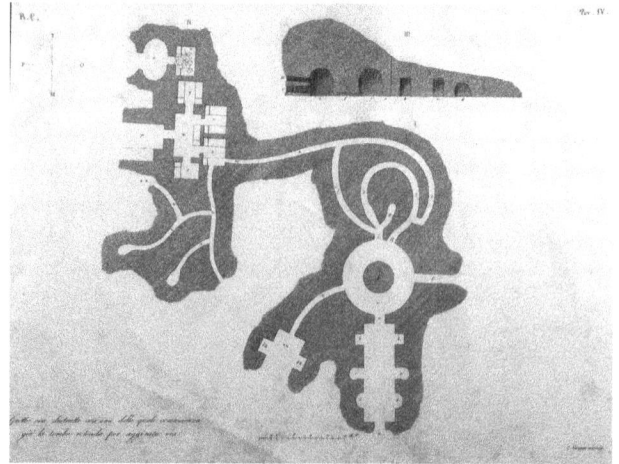

Fig. 9. Map of the labyrinth at Poggio Gaiella, near Chiusi, probably the most famous in Etruria (after BRAUN, 1840)

Systems like the ones in Tuscania along the Maschiolo course, the Acquarella, or the Marta river, which divert and engineer surface streams into underground tunnels, or hypogeic reservoirs such as the one at casale Lognazzo near Viterbo, all these offer a striking correspondence with the ones at Petra, Gardaïa, or who knows how many other less well-known oases around the world, excavated to divert flood waters and trap them through underground porosity. And it really does not matter whether in each case the principal function of the tunnel is transportation or storage, or on the other hand to integrate surface water by recovering the infiltrations along the way: in both cases its purpose is carried out using poly-functional solutions, which answer in different ways to each different ecosystem where they operate, and which have the potential capacity of adapting themselves to changes, continuing to function even in changed conditions. The same is true of the *cuniculi* at Ara del Tufo, Castel Broco or Bottacce 2, whose mouth is closed with a wall that holds in the water, and transforms them into reservoirs. Reservoirs whose walls are indeed porous and not plastered thereby allow water to pass underground, enabling the storage of surplus water there, so it can be available at later times, once the tunnel is empty again. This is a useful way of balancing the effects of climate change, at our latitude: there may be a reduction in quantity of rainfall, or a concentration of it in shorter periods, alternating with even longer droughts. The *cuniculi* of Siena are worth mentioning here, and Fraccaro remembers the cleaning and maintenance works usually done each year in winter, that being the season when the waters were lower, since 'for the time needed by seepage', the effect of the autumn rains was not yet felt (FRACCARO, 1919, 204). The innumerable *cuniculi* found underground, the excavated canals which implement or modify the surface hydrographic network, the artificial outlets which regulate the lakes or the closed basins, if looked at in their totality, constitute an exceptional network for water management, a distribution which, as in a leaf or the roots of a plant, enables a kind of osmosis between above ground and underground, a series of exchanges which, thanks to gravity alone, are able to use the underground level as a facilitator and thus limit the excess as well as the shortage of water, and finally convert the availability of water, variable between the seasons, into a more constant supply. This network is an everlasting phenomenon, repeated over the centuries and whose origins have been lost in the passage of time, whose functioning cannot be explained except as the sum of innumerable small interventions, and a 'miracle' that is as unfathomable as they are apparent, each individually, all the different and ordinary products and results of everyday maintenance and use.

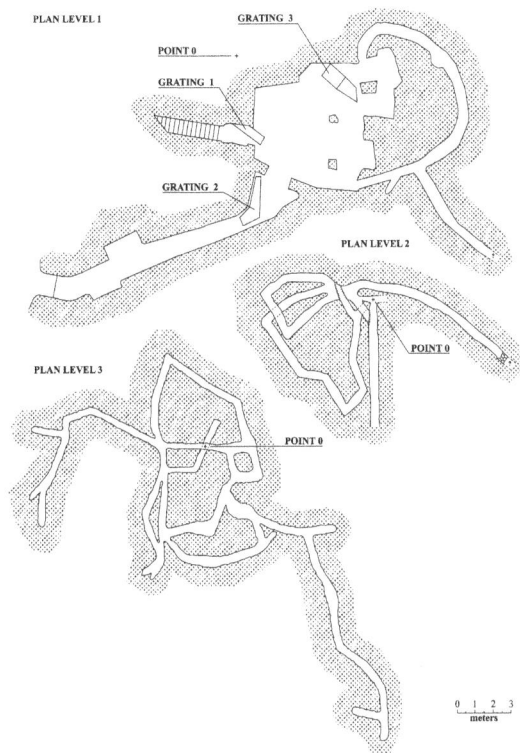

Fig. 10. Map of the so-called Tomba della Regina in Tuscania (after BERNARDONI, ORLANDI, 2002, 1-3, redrawn. Courtesy of Biblioteca Comunale di Tuscania)
(The three-layered labyrinth of *cuniculi* appears to be a *unicum* all over Etruria.)

And yet in many cases the functioning is actually unfathomable, or so it appears: in addition to the *cuniculi* of San Potente or Cunicchio, in which water capture is easily detectable as the main function, but the principle by which the water is collected is not evident at all, there are many other examples which evidently are nothing but portions of a wider system, and yet are still able to function. This is the case, for example, with Scalette, in which a very short portion of a *cuniculus* is still intercepting water through cracks in the rock, and feeding a beautiful fountain not far away. Of the bigger system, which originally took the water from the Marta River about a kilometre upstream, probably to irrigate further down, nothing is left but a small bridge where the canal passed over a ditch, and pieces of tunnel silted up but still aligned by both direction and elevation. The still ongoing

feeding of the fountain shows clearly how each system is in fact composed by many poly-functional integrated and potentially independent simple elements.

The miracle that fills up the reservoirs and gives water during the dry season cannot be understood except through acquaintance with the aqueduct of Acqua Vergine, one of Rome's aqueducts, which because of the many springs along its course looks like 'a huge *cuniculus* draining the underground waters of entire regions' (QUILICI, 1984, 68). We must also understand how the artificial outlets of closed lakes and basins were dug not only to lower the water level to create new agricultural land along the shore, but also to control the water to be used further downstream: a system well represented by the Nemi Lake, whose steep banks would have enabled drying just a few hectares of land for every metre of water lowered, and where the lower Ariccia crater, drained in turn by a similar system, represents the recipient water basin (CASTELLANI, 1999, 75-92). Or, in the end, we must understand the difficulties encountered by the Israeli government, during 1960, experimenting with the storage of water underground during the wet season by trying to reverse the flow within the vertical captation wells. This was conceptually a good idea which found its greatest obstacle in the wells' very low exchange surface (HARPAZ, 1965, 43-51).

The picture we get is one of an extremely wide and articulated reclamation effort which, from the *cuniculi* of just a few metres long and sparse around the country up to the huge lake regulation works, is composed of many structures, each multi-functional and complementary to each other. These solutions, although on different scales, are all based on the same concept: the continuity of the water cycle, its above ground and underground complementarity, the knowledge of how the precarious balance of the whole is based on each modest detail, and how, on the other hand, it is possible to intervene with these details, and modify the ecosystem in order to make it livable and productive, without provoking its collapse.

More than with morphological analogies, which change with crossed geology, and more than the fact that in many cases excavation techniques are fundamentally the same - involving the extraordinary consequence that still nowadays someone, somewhere in the world is able to reproduce the same achievements as in our past - the closeness between the Etruscan *cuniculi*, the Foggara and the drainage tunnels of other countries has probably to be researched within this general vision, in this capacity that humans had or still have of observing and adapting to natural dynamics so as to exploit them to our own advantage. Maybe more than in captation dynamics - which in oases are due principally to hidden rains, whereas the *cuniculi* not only rely on greater water availability, but in some cases this is not even the main function for which the tunnels were dug - the common thread to be searched for is the capacity to recognise and respect the limits imposed by environmental conditions, in the knowledge of how above ground and underground are two aspects of the same system. It is nice to think that this could be another reason why in the past the word *cuniculus*, specifically indicating narrow excavated passages underground, in the case of the hydraulic meaning has been used to indicate a canal excavated 'as much below as above ground'. (*Thesaurus Linguae Latinae*, entry *cuniculus*).

The Golden Valley

Independently from its function and chronology, each *cuniculus* tells a story of reclaimed land, of cultivation, of richness and improvement of environmental conditions. Those that are still functioning, thanks to the endeavours and efforts of those who excavated them, they repay the patience and time of those who maintained them as time went by: if not recognising their cultural value, they understood their function and adapted them to changing conditions or needs. The picture we see is one of an extraordinary long-term balance, the result of many tiny changes carried on through the ages, and the story of all the hands and the ideas above and beyond the gestures and knowledge necessary for its accomplishment. The Tuscania investigation, from this point of view, has surely been too short, only permitting us to see the outlines of a picture in which the empty spaces are still more than the filled-in ones. But it has also been, on the other hand, stimulating and innovative, thanks to the attention dedicated to the functional aspects as well, enabling us to extend the boundaries of a diagram that, until now, was mainly represented by archaeological research alone. It has also underlined the lack of source material, and the limited attention given to the subject: the fact that the 25 water systems identified within the Natural Park of Tuscania represent a density per square kilometre about three times higher than that of previously analysed areas (CAPONETTI, 2006, 19) does not mean a wealth in Tuscania but a deficiency of research coverage, which has never been organised into a systematic and complete study. In spite of the preliminary character of the research conducted until now, it is clear there is a need and urgency for a follow-up investigation leading to an inventory and cataloguing of the *cuniculi* of the Etruscan area, so as to demonstrate how central they in fact are to the study of the relationship between humans, technological civilization and nature. It is also necessary to deepen our knowledge of functioning dynamics, to actually verify the potentiality of a network that is much wider than imagined and to understand the dangers that threaten its functioning. At the same time it is necessary to pursue and widen the studies on landscape archeology, vital to fully understand the role these works had in taming the land, to depict the image of various agricultural sceneries that followed each other during the ages, and finally to enquire about the needs and necessities that were revealed each time by the excavation of a tunnel, needs and necessities that very likely will recur soon in the future.

The image we gained by the research in Tuscania is that of a landscape that was created by human hands and that needs humans to keep on living; but it is also that of an instrument which has already proved it can function in

the long run, and is able to sustain human activities even through climate change. This is an overview starting from its archaic origins with a legacy of the Neolithic; the subsequent works tell the story not only of successful reclamation and landscape modification, but also of perseverance and an accumulation of daily works, thereby stimulating consideration of the false contrast between conservation and the use of works that are our cultural heritage as well as functioning structures; this also underlines the paradox of modern territorial politics which creates an agriculture forced to live on subsidies, and which needs higher and higher costs for the management of environmental emergencies and hydro-geologic displacement. This is a panorama as invisible as it is determining, an underground landscape which reflects itself in the agricultural arrangements of the territory, and which in turn needs those same arrangements to be there, in order to function. From this point of view, the research has shown how the Park, and in general the environment around Tuscania, is in good working condition: the fact that in half the cases the *cuniculi* are still functioning is certainly due to continuing maintenance works, but also to the fact of their origins within an agro-ecosystem based on extensive agriculture, in which the disturbance brought about by the agricultural revolution that in the last fifty years has introduced the use of tractors and chemicals, that has substituted self-consumption with its specialized production, has been felt less than in other cases. It is not accidental that the Park is not presented for its naturalistic value but rather as an example of agricultural landscape derived from a balanced integration of its natural elements with human sustainable interventions. This makes it look even stranger that most of the lands where the *cuniculi* are found, due to their low workability with mechanical means, are today assigned to the margins of the agricultural system, arriving at the paradox of a situation in which the problem is not the conservation of works close to 3,000 years old, nor even the maintenance of their functionality, but how to manage to include the use of the water they produce - at zero cost - within a modern economic system.

The *cuniculi* ultimately are not just a remote legacy, although still working, of our ancestors' achievements, but also and above all a record of the links and relationships between resources and landscape, between productive activities and historic social processes, and a reminder of how these relationships reflect and form the agricultural landscape. Stuck 20 metres underground, with our feet wet in water and stuck in a tunnel no wider than our shoulders, we are reminded of 'the decline, not of the "Etruscan people", but of an arrangement of agricultural production and exchange' (TORELLI, 1981, 23) which was at the base of the environmental collapse of Etruria of that age. We can only wish that in a time like ours, confronted with questions on climate change, the hydro-geological imbalance, the agricultural crisis and energy issues, we will be able to remember and keep under consideration the value of small and insignificant works spread out around the country, and to imagine and establish a new and lasting Golden Valley.

Bibliography:

BAGNASCO GIANNI, G., 1996
La gestione delle acque in area etrusca: il caso di Tarquinia [in:] Acque interne: uso e gestione di una risorsa, Milan

BARKER, G., 1987
Archeologia del paesaggio ed agricoltura etrusca [in:] L'alimentazione nel mondo antico, Roma, vol. 2, 17-32

BERNARDONI, F. and ORLANDI, C., 2002
Progetto Tomba della Regina – Relazione sull'attività di studio e rilievo 2001-2002, self-published (Modena)

BIZZARRI, C., 1991
Cunicoli di drenaggio ad Orvieto [in:] Etruschi maestri di idraulica, Perugia

BODON, G. et al., 1994
Utilitas necessaria: sistemi idraulici nell'Italia romana, Rome

BRAUN, E., 1840
Il laberinto di Porsenna comparato coi sepolcri di Poggio-Gajella ultimamente dissotterrati nell'Agro Clusino, Rome

CAPONETTI, L., 2006
I cunicoli di Tuscania: un nuovo approccio per una indagine territoriale [in:] Analecta Romana Instituti Danici, XXXII, Rome, 7-26

BONI, P. et al., 1988
Carta idrogeologica del territorio della Regione Lazio, Rome

CASTELLANI, V., 1999
Civiltà dell'acqua, Rome

CUSTODIO, E. , LLAMAS, M. R., 2005
Idrologia sotterranea, Palermo [original edition: *Hidrologia subterrànea*, Barcelona, 1996]

DEL PELO PARDI, T., 1969
I cunicoli del Lazio, Rome

D'ERCOLE, V., DI GENNARO, F., GUIDI, A., 2002a
Appartenenza etnica e complessità sociale in Italia centrale: l'esame di situazioni territoriali diverse [in:] Primi popoli d'Europa – Proposte e riflessioni sulle origini della civiltà nell'Europa mediterranea (Palermo, 14-16 ottobre 1994 e Baeza 18-20 dicembre 1995), 127-36

D'ERCOLE, V., DI GENNARO, F., GUIDI, A., 2002b
Valore e limiti dei dati archeologici nella definizione delle linee di sviluppo delle comunità protostoriche dell'Italia centrale [in:] Primi popoli d'Europa – Proposte e riflessioni sulle origini della civiltà nell'Europa mediterranea (Palermo, 14-16 ottobre 1994 e Baeza 18-20 dicembre 1995), 111-26

FELICI, A., CAPPA, G. AND CAPPA, E., 1994
Il sistema ipogeo di alimentazione dell'acqua sacra al tempio di Giunone Curite [in:] Informazioni, III, n. 11

FIOCCHI NICOLAI, V., 1988
I cimiteri paleocristiani del Lazio, Città del Vaticano

FRACCARO, P., 1919
Di alcuni antichissimi lavori idraulici di Roma e della Campagna [in:] Bollettino della Società Geografica Italiana, serie V, vol. 8, 186-215

GIANNOTTI, F., 1606
Storia di Tuscania scritta da Francesco Giannotti nel 1500 (manuscript, Tuscania, Public Library)

GIULIANI, R., 2006
Genesi e sviluppo dei nuclei costitutivi del cimitero di Priscilla [in:] Origine delle catacombe romane, Città del Vaticano, 2006, 162-75.

HARPAZ, Y., 1965
Field experiment in recharge and mixing though wells [in:] Underground water storage study, XVII, 3-54, Tel Aviv

JUDSON, S. , KAHANE, A., 1963
Underground drainageways in Southern Etruria and Northern Latium, Papers of the British School at Rome, 31, 74-99

LAUREANO, P., 1995
La piramide rovesciata, Torino

LIVERANI, M., 1998
Uruk, la prima città, Bari

PERGOLA, P., 1997
Le catacombe romane. Storia e topografia, Roma

PERGOLA, P., 2006
Gli ipogei all'origine della catacomba di Domitilla: una rilettura [in:] Origine delle catacombe romane, Città del Vaticano, 2006, 176-84

QUILICI GIGLI, S., 1983
Sistema di cunicoli nel territorio tra Velletri e Cisterna [in:] Archeologia laziale, 5/83, 112-23

QUILICI, L., 1987
Il sistema di captazione delle sorgenti [in:] Il trionfo dell'acqua - Gli antichi acquedotti di Roma:problemi di conoscenza, conservazione e tutela (Roma, 29/30 ottobre 1987), 47-51

RASTRELLI, A., 2000
Chiusi Etrusca, Chiusi

RAVELLI, F. , HOWARTH, P., 1988

I cunicoli etrusco-latini: tunnel per la captazione di acqua pura [in:] Irrigazione e drenaggio, 35, vol. 1, 57-70

SIGNORELLI, M., 1966
Sui sentieri dei Lucumoni etruschi, Viterbo

SPERA, L., 2004
Il complesso di Pretestato sulla Via Appia, Città del Vaticano

Thesaurus Linguae Latinae, 1900
Lipsia

TOLOTTI, F., 1978
Origine e sviluppo delle escavazioni del cimitero di Pretestato [in:] Atti del IX congresso di archeologia cristiana, Roma, 159-87

TOLOTTI, F., 1980
Influenza delle opere idrauliche sull'origine delle catacombe [in:] *Rivista di Archeologia Cristiana,* LVI, 7-48, Roma

TORELLI, M., 1983
Storia degli Etruschi, Bari

WARD PERKINS, J. B., 1962
Etruscan engineering: road building, water supply and drainage [in:] Hommage à Albert Grenier, Bruxelles, 1642-3

THE SOCIAL ROLE OF ROMAN BATHS IN THE PROVINCE OF MOESIA SUPERIOR[1]

Marko A. Janković
Department of Archaeology,
Faculty of Philosophy,
University of Belgrade

Key Words: Moesia Superior, baths, bathing, identity, social interaction

Introduction

Bathing facilities were discovered in almost every part of the former Roman Empire, and very few other objects are so clearly and unquestionably marked as Roman. Baths were often used as one of the diagnostic tools for uncovering cultural changes that occurred in the non-Roman societies after their annexation to the Empire (the process labeled as Romanization in a vast body of literature). Baths, among other structures (such as theaters, amphitheaters, etc.) have been observed as the indicators of social changes and their construction has been recognized as the certain sign of the acceptance of the Roman culture as superior and more civilized. In such evolutionary imagined social transformation, it is believed that the non-Roman societies must have accepted the Roman culture as a naturally better strategy of living. Nevertheless, in the last two decades more than a few significant papers have been published on the subject of Romanization, dealing with different aspects of this narrative (MILLET, 1990; WOOLF, 1998; HINGLEY, 2000; WEBSTER, 2001; REVELL, 2008). Some of these works stress the limitations and dead-ends of Romanization theory (HINGLEY, 2000), some are dealing with mechanisms of social changes different from the traditional (WOOLF, 1998) and yet others are focused on the issues of maintenance of Roman identity once when it was created (REVELL, 2008). All of the critics agree that the mere presence of a material culture labeled as Roman does not necessarily mean that social change was taking place. It has also been suggested that such Roman objects (small artifacts or monumental architecture) could be used for maintenance of different identities (REVELL, 2008, 3).

This paper treats baths within a wide chronological framework, and they are observed in different historical and social contexts. The Roman baths in the territory of the province of Moesia Superior were built over a long period of time and for different purposes, so it is very important to observe them in their own context rather than through a precise catalogue that could give us accurate data of their architectural type, geographical position or dimensions. The basic information will be provided for each bath (where possible), but the main focus will remain on their role in everyday life and social interactions. In order to outline such a role, I had to stress features of the baths that I believe were indicative. First of all, I had to separate the group of early constructed baths from the others, due to a different historic context in which they were built. Within each of the groups, attention was paid to different contexts in which baths were erected. It was important because these different contexts could explain different needs (e.g. civilian / military) and therefore different characteristics of the baths. The last important criterion was the presence of social activities other than bathing.

The Roman baths in the province of Moesia Superior were built almost throughout the period of the Roman presence in this territory. A few objects were dated to the 1st and 2nd centuries AD, while the majority of the buildings were erected during the 3rd and used until the end of the 4th century (Fig. 1). This time framework exceeds the duration of the province of Moesia Superior (which lasted for the period AD 86–273), but it was important to track the continuation of the usage of these objects. This approach offers valuable information on their longevity and popularity among the local population. Some of these objects were "transformed" and used as sacred places long after people stopped using them. Naissus, Nerodimlje or Egeta baths are just some of the many examples where people from the Middle Ages used the Roman baths as burial grounds.

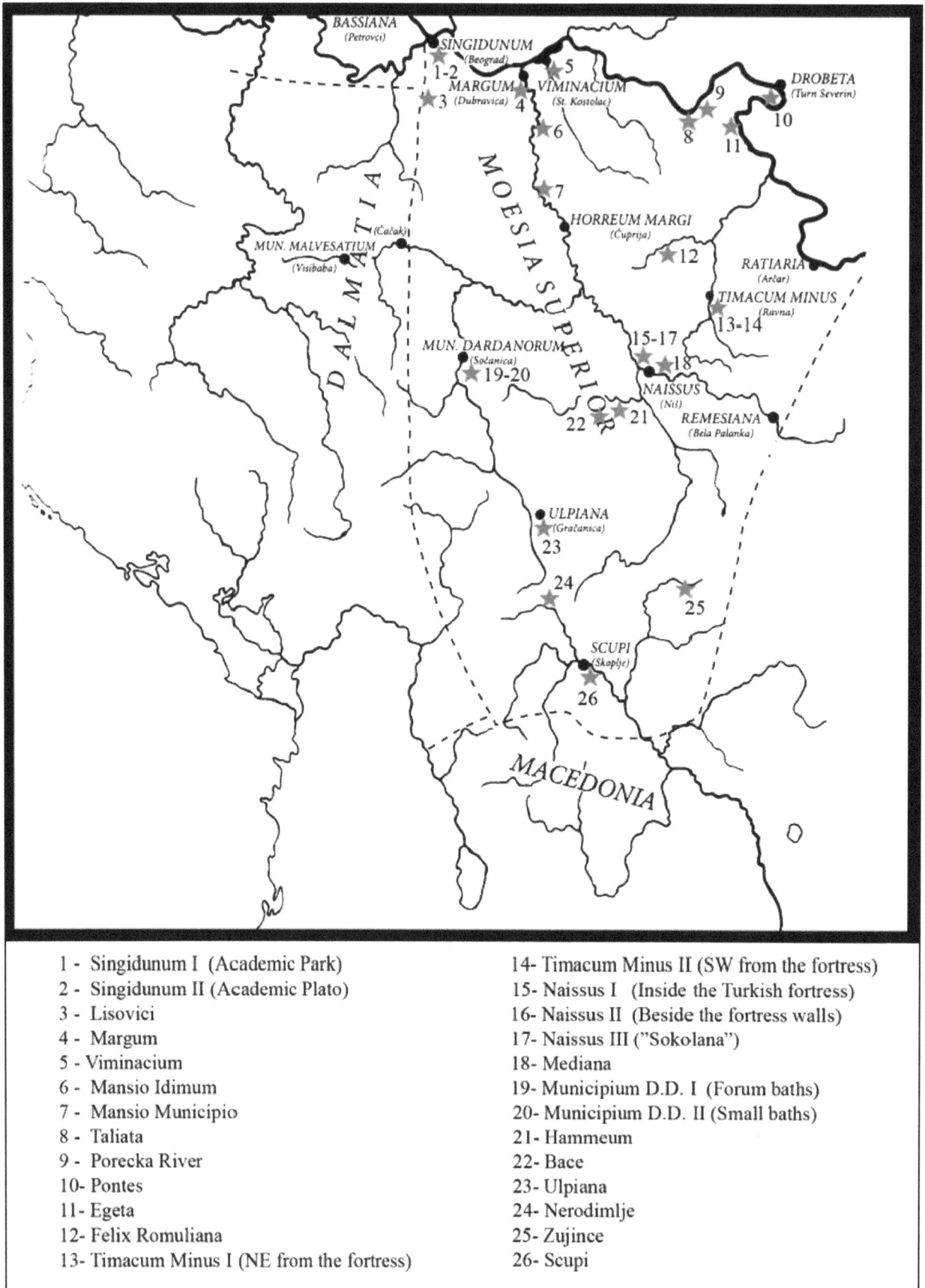

Fig. 1 Map of the distribution of baths in the province of Moesia Superior

1 - Singidunum I (Academic Park)
2 - Singidunum II (Academic Plato)
3 - Lisovici
4 - Margum
5 - Viminacium
6 - Mansio Idimum
7 - Mansio Municipio
8 - Taliata
9 - Porecka River
10- Pontes
11- Egeta
12- Felix Romuliana
13- Timacum Minus I (NE from the fortress)

14- Timacum Minus II (SW from the fortress)
15- Naissus I (Inside the Turkish fortress)
16- Naissus II (Beside the fortress walls)
17- Naissus III ("Sokolana")
18- Mediana
19- Municipium D.D. I (Forum baths)
20- Municipium D.D. II (Small baths)
21- Hammeum
22- Bace
23- Ulpiana
24- Nerodimlje
25- Zujince
26- Scupi

Nevertheless, the timeline of the paper ends with the abandonment of the baths for their original purpose, which could be set for most of the baths at the end of the 4th or the beginning of the 5th century. The only exception was confirmed at Nerodimlje, where the baths were adapted and used during the 6th century (LAZIĆ, 2001, 279).

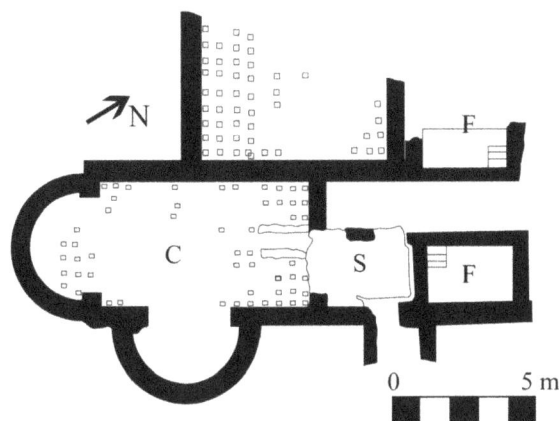

Fig. 2 Baths at Academic Park (after Bojović, 1977, 7; modified by M. Janković)

Due to the quality of the excavations conducted and published reports, it was not possible to give a definite answer to all the questions asked in this paper, but the intention was only to ask as many proper questions as possible and to try opening the debate on some issues. This is also the first time that the baths from Moesia Superior are published jointly, giving an opportunity to other researchers to compare, check and even to supplement their own researches. Today, we have several monographs (DE LAINE, 1988; NIELSEN, 1990; YEGÜL, 1992; FAGAN, 1999) and symposia reports (DE LAINE, JOHNSON, eds, 1999) on baths, but none of these volumes ever discusses the baths from this part of the Roman Empire. This lack of a synthetic work, especially in languages other than local, surely influences such a situation, and I hope that I will at least shed some light on the subject.

Early Baths of Moesia Superior

The earliest dated baths were built at Viminacium, a legionary fortress and capital of the province. Baths were set in the civil settlement and dated to the 1st century AD (KONDIĆ, ZOTOVIĆ 1974, 97). The researchers of these baths dated the object by comparing the material registered inside the baths with those found in the other settlement objects in the vicinity. Later excavations gave us plenty of data securing precise dating for usage of the baths until the end of the 4th century (MILOVANOVIĆ, 2008). Most of the bath area was heated by exterior furnaces, while the rooms for cold bathing had a glass ceiling or walls, judging by the great amount of glass fragments; the interior was decorated with mosaic floors and polychrome frescoes on the walls. A great number of

personal objects, such as hairpins, fibulae or arm rings, were found in the dressing room.

The other "early" baths were erected during the 2nd century AD. All of these were placed at important early Roman sites such as forts (Margum, Timacum Minus I, and Egeta) or mining centers (Municipium D.D., Timacum Minus I, Lisovići). It is very hard to discern much regularity among these baths. They are very different in extent, building techniques and interior decoration. Some of them were never completely excavated so we don't have all the data for reconstruction of the number of rooms or decoration. All of them had basic rooms for cold and hot bathing, but not a single room for any other social activity has been confirmed yet. Early baths were lavishly decorated with mosaics, frescoes and even marble and stone sculptures.

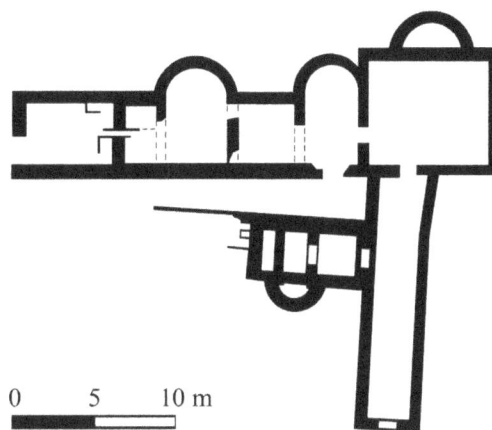

Fig. 3 Baths at Taliata (after PETROVIĆ, VASIĆ, 1996, 26; modified by M. Janković)

But there is one thing we can say with great certainty. We are able to identify the connection between the building of these earliest baths and the places of great importance – the Roman mines. All three different baths are built in three different areas where the Roman mining activity was confirmed: the Ibar valley, Kosmaj Mountains and the Timok valley. Municipium D.D. is the mining center of the southern part of the province – Dardania (50 km NW of Priština), and a number of slag damp sites and mining facilities were uncovered in the valley of the river Ibar (DUŠANIĆ, 1977, 72). The site of Lisovići is placed in the Kosmaj Mt. hinterland (25 km SW of Belgrade), also a confirmed Roman mining site. Kosmaj Mt. is an area of well documented high mining activities. Refineries, damp sites, but also a military fort (Stojnik) and a necropolis were excavated during the last century on the slopes of Kosmaj Mt. (VELIČKOVIĆ, 1957; DUŠANIĆ, 2000; MERKEL 2007). In the eastern part of the province, a great area of ore exploitation was confirmed. One of the important centers of this area was located at Timacum Minus (the village of Ravna). The *cohorta Aurelia Dardanorum II* was stationed at the auxiliary fort of Timacum Minus, and its main task was to ensure caravans to and from these mining areas (DUŠANIĆ, 2000, 349). The other early baths (Margum, Egeta) were erected near these areas, but the quality of available data does not allow us specific conclusions on their mutual relations.

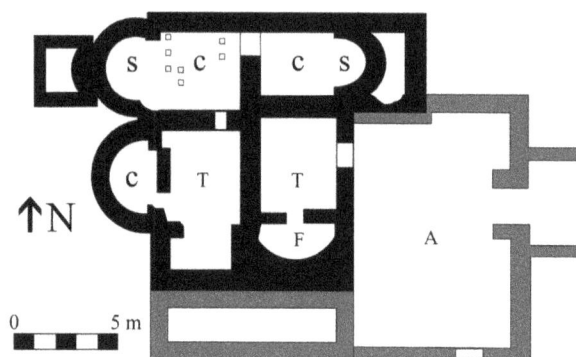

Fig. 4 Baths I at Timacum Minus (after PETROVIĆ, JOVANOVIĆ, 1997, 24; modified by M. Janković et al., 2005, 16)

All of these objects are designed differently and, unfortunately, some of them were never completely excavated – like Egeta (PETROVIĆ, 1987), Lisovići or Municipium D.D. – so there is a problem of lack of data, which is important for precise comparison of their interiors or the building techniques applied. The baths at Lisovići are probably the largest, and consisted of a great number of heated rooms; they were probably just a part of a more major structure (VELIČKOVIĆ, 1957, 377). The excavation reports say nothing about the material recovered there, and it remains only an assumption that some of the rooms were used for social activities other than bathing. The baths at Viminacium, Egeta and Municipium D.D. were decorated by fresco technique, and a large body of polychrome fragments was discovered; it was even possible to reconstruct some of the motifs as geometric and floral decorations in Viminacium and Egeta, or a motif of a swamp bird at Municipium D.D. The floors of the baths were covered with marble slabs at Viminacium, while mosaics were used at both Egeta and Viminacium. In both baths, Egeta and Lisovići, a stone female figure was found in the interior.

Late Baths of Moesia Superior

As mentioned above, the majority of the baths were erected during the late Roman period. There are 21 confirmed bath-houses all over the province, built in different landscapes and settled in different social, geographical and historical contexts. Thanks to the greater sample, it is possible to draw much more valuable data from these objects than the earlier ones. Therefore, it is not possible to observe them as a whole, but different groups of objects can be proposed based mainly on their social context. The main division – between civil and military baths – is the most common in all Roman provinces, but we shall see that it is not always applicable

for the province of Moesia Superior. It is very hard to draw a straight line between these two groups, simply because there is no single bath registered on the perimeter of a military fort, neither legionary nor auxiliary.

Military baths form one of the inevitable subjects for all the researches dealing with Roman baths, and some of these observations are very helpful as analogies for the province of Moesia Superior (NIELSEN, 1990; REVELL, 2007). Speaking of military baths, authors separate the baths of legionary forts from those registered in smaller auxiliary forts. Nielsen is working with a much broader sample, and she makes a difference between legionary and *castellum* baths (NIELSEN, 1990, 77). Beside others, she points out that the legionary forts located inside the forts, and most smaller *castellum* baths, are located outside the perimeter of forts. On the other hand, Revell focuses her research on baths within the sole province of Britannia, again separating two groups: legionary and auxiliary baths (REVELL, 2007, 231). In the territory this paper is dealing with, the main problem is the lack of legionary baths. Both legionary forts (Viminacium and Singidunum) showed no trace of baths inside their walls yet. Nevertheless, there is a tiny possibility that some of the "civilian" baths from the vicinity may have been used for military purposes.

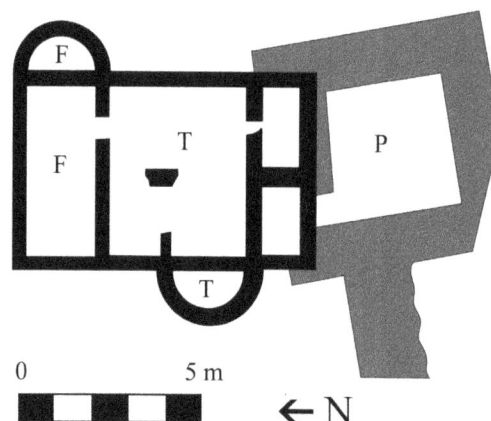

Fig. 5 Baths at Porecka Reka (after PETROVIĆ, 1984, 289; modified by M. Janković)

The civil settlements at both centers had one (Viminacium) or two (Singidunum) bathing objects built very near to military forts. It is most likely that civilians and soldiers used them separately. Both the baths at Viminacium and Singidunum, the Academic Park baths (BOJOVIĆ, 1977), are very large in their extent (at least in scale proper for Moesia Superior). Aside from basic bathing facilities, they contained rooms for other activities. First of all, the baths in Academic Park are a rare example containing a sweat room (Fig. 2). One of the rooms, small in dimension (3, 15 x 2, 20 m) and connected with the *caldarium*, had its own furnace, a hypocaust system and a system for wall heating made of *tegulae mamatae*. All these characteristics led the original researchers to mark this facility as a sweat room

(BOJOVIĆ, 1977, 20). Also, both baths had spacious *apoditeria*, which were not confirmed for most of auxiliary baths – Taliata (Fig. 3), Egeta, Pontes, Timacum Minus II, Naissus, Ulpiana. The only exceptions are the baths I at Timacum Minus (Fig. 4) where an *apodyterium* was added in the second phase of usage (PETROVIĆ, 1984).

Unfortunately, most of the auxiliary baths have never been completely excavated, so we do not have all the relevant data. There are six sites where baths are definitely connected to military forts and three of them are placed on the Danube *Limes* – Taliata (VUČKOVIĆ-TODOROVIĆ, 1969), Pontes (PETROVIĆ, VASIĆ 1996), Egeta – while the other three were constructed in the vicinity of large forts inside the province, in the southern part known as Dardania – Timacum Minus (PETROVIĆ, JOVANOVIĆ 1997), Naissus (PETROVIĆ, 1999) and Ulpiana (PAROVIĆ-PEŠIKAN, 1981). The baths at Porečka River Confluence probably belong to this group, but the problem is that there is no fort in the vicinity. Actually, the baths were erected on top of the former fort (Fig. 5). During the protective excavations (1967–70) several different objects were registered. The character of these objects (fortified granary, warehouses) led to the assumption that this was the collection center of the Roman army. At the time when the baths were operating, the fort was abandoned and partially damaged, but the watchtowers and fortified wall built perpendicularly to the Porečka River were still in use (PETROVIĆ, 1984, 290). Its position in regard to the Roman *Limes* and the connection to military affairs classify it as auxiliary baths. The only confirmed rooms

are marked as bathing facilities, and there is no single evidence that some part of this baths was ever used for any other social activity. Such a state is completely in accord with the research results of previous researchers. Still, we have to bear in mind that there is an obvious lack of data, so this conclusion may be refuted in the future.

As mentioned above, the group of civilian baths is much larger than the group of military facilities. It is possible to discern subgroups within the group of civilian baths, based mainly on their relation to the rest of the settlement or the type of "settlement" where they were erected. These subgroups are used simply for better comparison between them and easier manipulation. If we are to accept the original dates, all of these baths were erected or used throughout the late Roman period (3rd–4th centuries). The first group of civilian baths covers the bathing facilities uncovered inside the civil settlements, with no other confirmed connection to structures around them – Singidunum-Academic Plato (BOJOVIĆ, 1977), Felix Romuliana (STOJKOVIĆ-PAVELKA, 2004), Naissus III (Petrović 1999), Municipium D.D. II (ČERŠKOV, 1970), Zujince (TOMOVIĆ et al., 2005), Hammeum (MILOŠEVIĆ, 1999) and Bace (JORDOVIĆ, 1999). Three baths were grouped together based on the fact that they were uncovered as part of luxurious palaces or villas – Mediana (LATKOVIĆ et al., 1979; PETROVIĆ, 1993), Nerodimlje (LAZIĆ, 2001) and Scupi (KORAČEVIĆ, 1985). Finally, the last group consists of two facilities which were a part of road stations – Mansio Idimum, Mansio Municipio (VASIĆ, MILOŠEVIĆ, 2000).

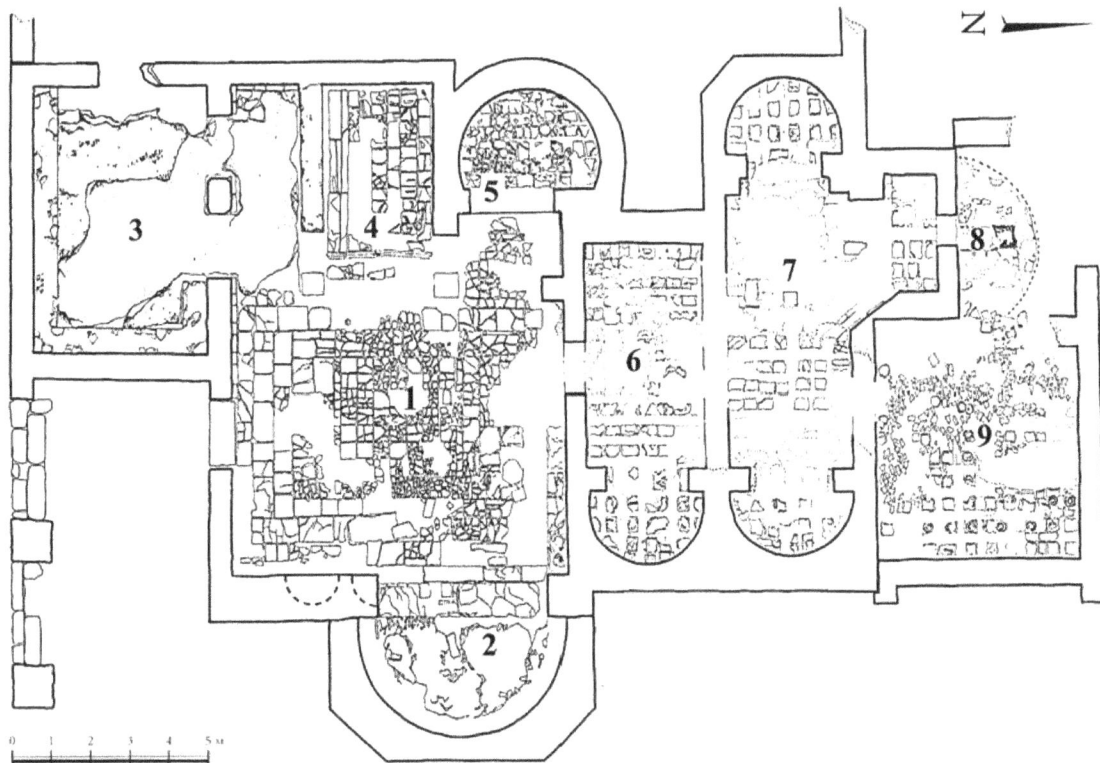

Fig. 6 Baths at Fekix Romuliana (after ČANAK-MEDIĆ , STOJKOVIĆ-PAVELKA, 2010, 100)

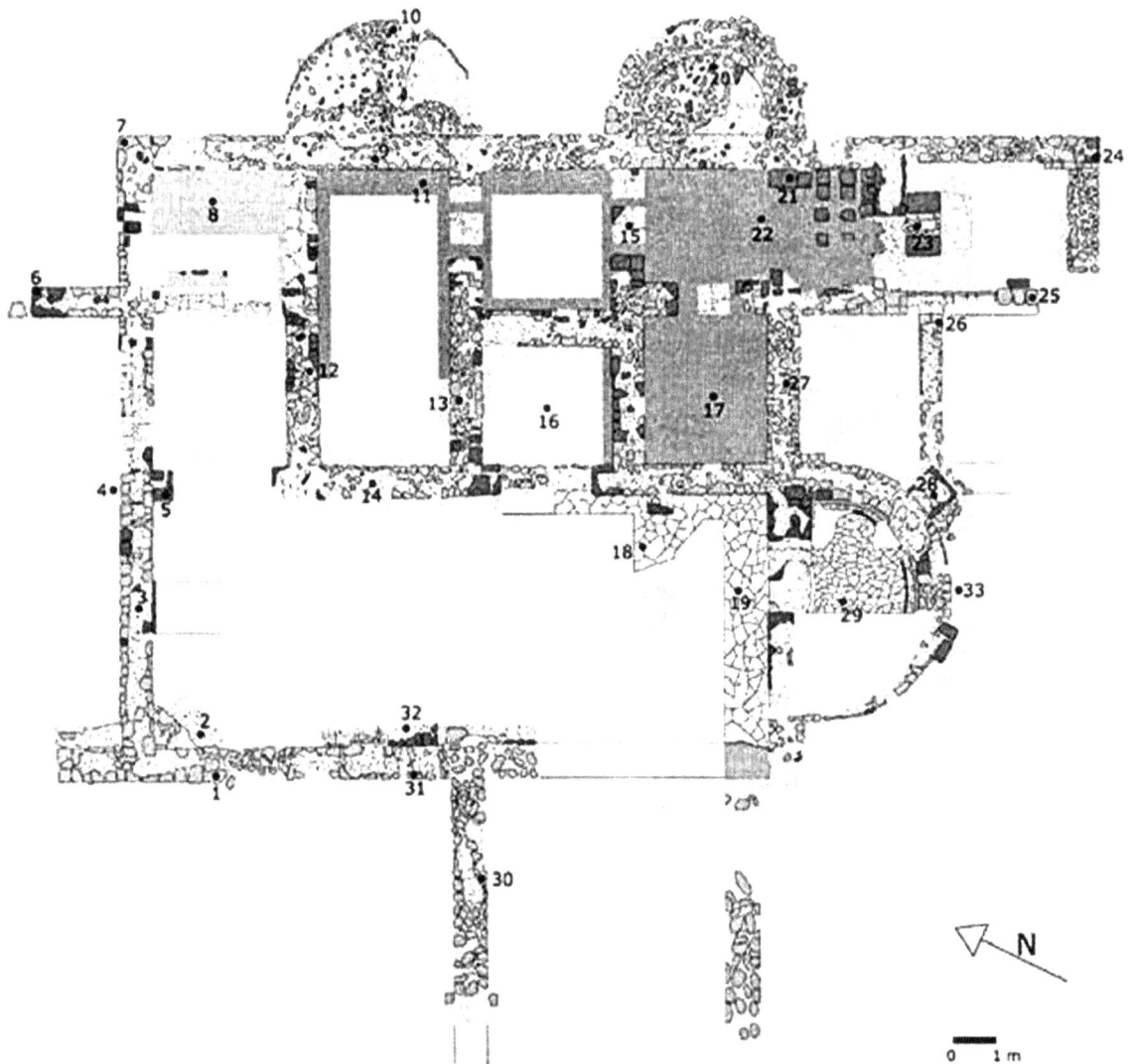

Fig. 7 Baths at Zujince (after TOMOVIĆ, BULATOVIĆ, KAPURAN, 2005, 322)

The baths discovered inside the settlements differ in size, number of facilities and type of architecture. While the baths at Municipium D.D. and Naissus III had only two small rooms, the baths of Felix Romuliana (Fig. 6) and Zujince (Fig. 7) were pretty complex and had numerous interconnected rooms. By contrast, the bathing facilities in Bace, Naissus III and Municipium D.D. were arranged in a row, but Felix Romuliana, Hammeum and Zujince showed a different ring-type of organization. Thermal facilities at Hammeum, Singidunum and Naissus III were never completely excavated due to different technical issues. But it is interesting that most of these baths were never decorated from the inside, the only exception being the facility at Felix Romuliana, but these baths were placed inside the settlement around the palace, so it was very likely part of a broader complex of the luxurious palace (ČANAK-MEDIĆ, STOJKOVIĆ-PAVELKA, 2010, 102). This group of baths in general shows no great difference in comparison to the military baths.

The group of three luxurious baths consists of baths discovered in different parts of the province. Two of them were recognized as a part of larger architectural complexes, while the baths at Nerodimlje were erected as a single luxurious structure. The baths of Scupi were integrated into larger buildings, denoted as palaces, while the baths of Mediana were erected by the luxurious villa located on the road from Naissus to Serdica. Once again, the Scupi baths were only partially investigated, so all we know is that they were integrated into a larger luxurious building, lavishly decorated and heated with a hypocaust system (KORAČEVIĆ, 1985). On the other hand, Mediana presents a major Roman site with a long research tradition within Serbian Roman archaeology. Also, this site was connected with the figure of the Roman Emperor Constantine the Great, which ensured enough valuable data for the baths located at this site. The Mediana baths were built as a separate building, but connected to the villa by a narrow hallway.

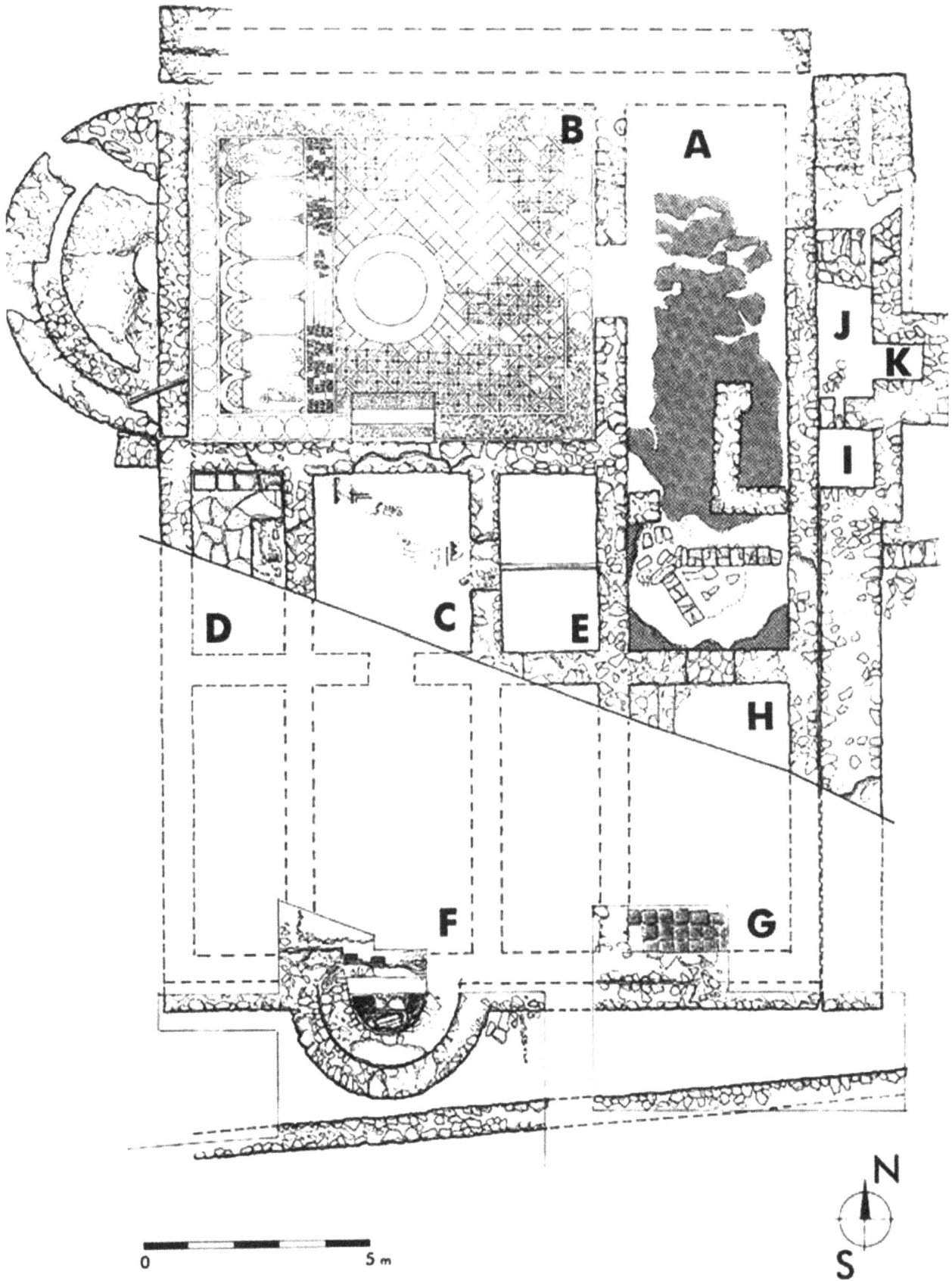

Fig. 8 Baths at Nerodimlje (after LAZIĆ, 2001, 253)

The baths are large, ring-type objects with several different facilities. At least two phases were confirmed and the baths are dated, as the rest of the site, as 4th century. The name of Mediana was attached to several 4th century Roman Emperors, above all with the name of Constantine the Great (AD 305–37), so it is widely believed that Constantine, above all, deserves the name of the benefactor of the villa. One of the reasons is the luxuriously decorated villa constructed together with its baths. The floors were covered with a combination of marble tiles and mosaics with geometrical motifs, while the walls were covered with floral motif frescoes. All the facilities were confirmed as bathing rooms, and there was no space for any other rooms serving any social activity. One other advantage of the landscape of Mediana was the abundance of hot mineral springs, so there is an assumption that the water installation made for the villa and baths was derived from such springs. Another interesting piece of data came from a hoard found in the villa in 1972, containing a group of statues made of marble and porphyry. Most of the statues show the figures of Asclepius, Hygeia, Telesphorus and Heracles (JOVANOVIĆ, 1975, 57). The bronze railing decorated with Asclepius and Luna busts was yet another argument for the researchers to assume the existence of a small temple at Mediana, probably consecrated to the group of iatric deities (VASIĆ, 2004, 103). The connection of such deities with baths and water facilities was not unique. A similar relation could be confirmed at Osmakovo and Krupac near Pirot in SE Serbia (PETROVIĆ, 1966, 250).

The thermal object at Nerodimlje uncovered twelve rooms built during the 3rd and 4th centuries (Fig. 8). The baths were built as a single structure, but it was confirmed that it was probably surrounded by a stone wall. The facilities were used throughout the 6th century, when the last adaptation was made. Apart from basic facilities (*caldarium*, *tepidarium*, *frigidarium*, *apodyterium*), it is very important that it had a *palestra*, which was very unusual for this territory. The *palestra* was lavishly decorated, this time with figures and texts. In the western part of the room the mosaic floor shows seven arcades, and in all of them seven figures are presented together with inscriptions identifying them as the seven wise men of antiquity (LAZIĆ, 2001). Below these figures, the inscriptions with famous sentences were placed. The rest of the *palestra* was decorated with mosaics in a square patchwork of rosetta and cross motifs. The center of the room was embellished with a luxurious fountain covered in mosaic on both interior and exterior. The *frigidarium* also contained mosaics with figures of nude females (probably nymphs) and a head of a deity with a wreath or crown in his hand (LAZIĆ, 2001, 256). Small pieces of mosaics were discovered in the rest of this room, and also in the *apodyterium*. As expected, at least by comparison with other late Roman baths from the area, analogies showed that mosaics were made in the 6th century, long after the building of the baths. This place was used as a burial ground later in the Middle Ages.

Although the baths at Mansio Idimum were not completely excavated, this is probably the best preserved and documented site of this kind in the region of Moesia Superior (VASIĆ, MILOŠEVIĆ, 2000). It was the first time that the contexts of small finds were published properly. The researchers were conscientious about each small find and allowed us to reconstruct its position inside the baths. The baths contained facilities for bathing, but it is important that other social activities were confirmed within certain rooms. First of all, it is the first time that some of the facilities were marked as "fun-room". A wide rectangular room was decorated with frescoes and had glass windows. In the extension to this room, the original researchers assumed a walled garden (VASIĆ, MILOŠEVIĆ, 2000, 55). These baths also had a sweat room – a small rectangular room with a hypocaust and a system for wall heating, and a wide rectangular room that was used as an *apodyterium* or *palestra* (Fig. 9). Among other small finds, tools for gardening and maintenance were found (scissors, knives, pickaxe, chisels, etc.), but probably the most important finds concerning social activities other than bathing were gaming counters. These kinds of objects are not uncommon in Moesia Superior, and they are confirmed in dozens of different Roman sites (JANKOVIĆ, 2010, 56). Still, this is a unique context and it should illustrate active social interaction from attending baths.

Another set of baths built at a road station were found at Mansio Municipio, not far from the previous site. These baths were partially excavated and only a couple of heated rooms were confirmed, with numerous pottery fragments, bronze coins and a fibula. The baths were dated as 4th century and probably had two phases of usage until they were burned in a fire (VASIĆ, MILOŠEVIĆ, 2000, 145).

Discussion

The thermal facilities described above are not subjected to a detailed catalogue, but I rather preferred to describe their contexts and those characteristics that I consider valid for discussing the social activities. Due to the late date of conquering the province of Moesia Superior and an even later date for beginning building the baths, the baths erected in this province were previously fully developed. A transition from hygienic object to places of social interaction in everyday routine already occurred at the beginning of the 1st century AD (DE LAINE, 1992; REVELL, 2007). Nevertheless, there are not many traces of such activities in the baths of Moesia Superior, and reasons for that might be not only archaeological but also technical. I mentioned a few times that the majority of these sites were only partially excavated and fragmentarily published. However, it was possible to reconstruct some facilities and activities which had nothing to do with maintaining hygiene. An *apodyteria* was confirmed in several different bath-houses, but there were few attempts to explain its purpose (VASIĆ, MILOŠEVIĆ, 2000; CUNJAK, 1996). At the baths of Viminacium, numerous hairpins and arm rings were discovered, together with Roman coins, and researchers concluded that it should be denoted as a dressing room (MILOVANOVIĆ, 2008). This is also a rare case where small finds were related to actual facilities. These rooms were commonly connected not only with changing clothing but also as facilities for non-bathing social activities (REVELL, 2007, 235). Still, little effort was put to interpreting these facilities. All we can do now is assume that these large unheated and non-bathing rooms were the central space for social interaction within the baths. Lack of these facilities is obvious at military auxiliary baths, and we can suppose that these baths with only basic bathing rooms were not used as other larger baths for numerous activities, but only for basic hygiene. From this point of view, the baths of Mansio Idimum are important in many ways. First of all, these baths are the only ones where original researchers tried to connect facilities and small finds found there with actual activities (VASIĆ, MILOŠEVIĆ, 2000). They assumed the existence of a "fun-room" and an open air garden based on the distribution of small finds and architectural design (an open-walled area with no confirmed floor).

The main purpose of these baths was to welcome travelers on the road, and it looks as if the bath designers exploited all possible advantages of the landscape. The other important finds of gaming counters testify to board games engagements. It is also an activity for which a special (not heated and certainly non-bathing) room was needed. Board games were very popular in the province of Roman Moesia, and many finds from more than thirty Roman sites indicate the popularity of nine-men's-morris and a *latrunculi* game (JANKOVIĆ, 2010, 60). Another find important for illustrating social activities was found at Timacum Minus (Baths II). A fragmented bone fife was discovered within the walls of these baths, and we can only assume the existence of at least one musician (PETROVIĆ, JANKOVIĆ, 1997, 99). Many other small finds like arm-rings, hairpins, fibulae, rings or various bronze keys, coins or plates were found all around the baths and they testify to the high level of activity inside the baths' walls. However, this cannot be used to illustrate particular social activity. Numerous finds of pottery fragments have been discovered, but only a small part of them (like *terra sigillata* pottery) was published, so it is not possible to conclude whether this pottery was used for oil, food, or something else (e.g. BOJOVIĆ, 1975).

The other aspect of this research dealt with the social and historical context of building and usage of Roman baths. The division into early and later Roman baths is not based simply on the historical dates of construction. Different circumstances led to baths' erection, and different mechanisms of usage were in place. It is my intention to deem the thermal facilities as objects used for the maintenance of Roman identity in a provincial context. Although the baths from both periods were architecturally very similar, it is my opinion that they were not used for exactly the same purpose. The earliest baths were constructed during the time of the first contact between the local population and the Roman conquerors, after the annexation. Together with other archaeological situations from the same period, these baths may be observed as part of the debate on cultural transformation, often picturesquely portrayed by the term "becoming Roman" (WOOLF, 1998; HINGLEY, 2005). On the other hand, most of the later baths were constructed in slightly different urban landscapes as part of larger settlements, well connected within the province. From that point of view, later baths may serve to explain a way of maintaining already created Roman identity, or simply a way of "being Roman". This inexact division does not mean that we can sharply separate different identities, but only that we must pay attention to different aspects observed in different contexts. Identities are not strict demarcations, identical at all times and in all circumstances, but fluid and changeable; and with that in mind that is how we should approach these objects (JONES, 1997; CASELLA, FOWLER, 2004; DIAZ-ANDREU et al., 2005; HINGLEY, 2005; REVELL, 2008).

Becoming Roman implies a certain change in everyday routine. It is something that people are not used to, but try to change, achieve and maintain. Models of behavior (in the sense of social norms), everyday routine and ways of interaction with others or through the usage of different political and social concepts are just a part of their image that people are trying to change. In other words, in order to become Roman, people embrace different strategies and create a whole new personal and collective identity. It is clear that these personal identities are not mutually equal, but we equalize them on a collective scale, and recognize them as Roman. So, the starting point is a personal and collective change in order to achieve different identities, to achieve a sort of balance between the old and new conditions, using different strategies. In this case it is a strategy of living in a whole new world – the Roman Empire (CURCHIN, 2004, 122; HINGLEY, 2005, 471).

On the other hand, being Roman implies that certain changes have already occurred in the past. If someone already became Roman, we should now try to reconstruct what mechanisms and strategies he is using to maintain his/her Roman identity. These changes and continuities are most visible in everyday routine: the way people use certain objects and also the facilities they frequent during their spare time, in this case in the baths (REVELL, 2008).

Material culture could not be seen as a particular form of Roman identity, but rather as a means through which usage identity is maintained; and the mere presence of certain material culture is not always a diagnostic feature for tracing cultural transformation. The social context of material culture and the way it has been used within new circumstances on the other hand may be helpful in such researches.

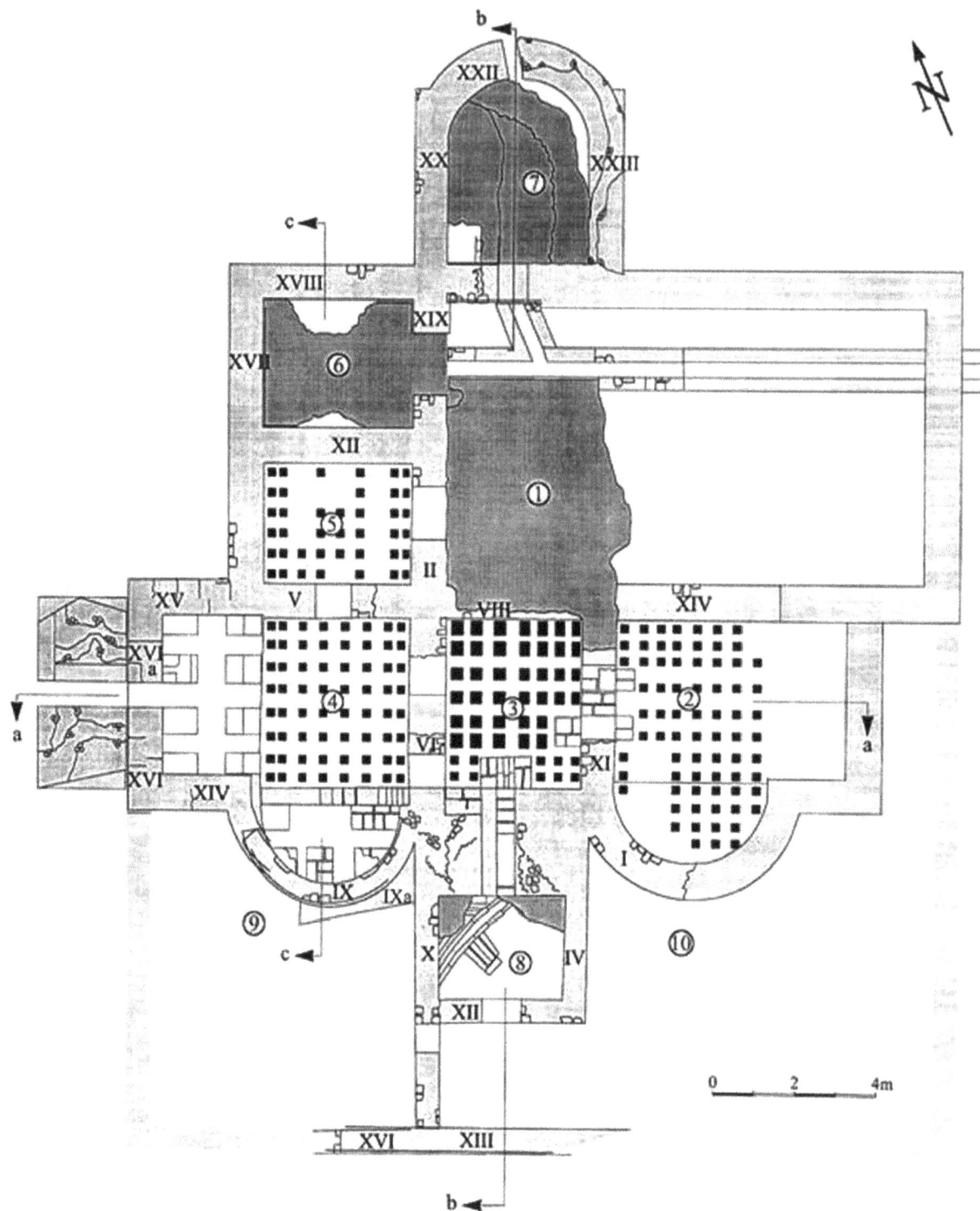

Fig. 9 Baths at Mansio Idimum (after VASIĆ, MILOŠEVIĆ, 2000, 52)

However, Roman baths are rather different in the sense that they were all erected after the historical date of occupation and establishment of the province, and their uniqueness is emphasized by the fact that there are no bathing facilities of any kind found in this area before the actual arrival of the Roman army and administration. As we saw earlier, the first baths were erected almost a century after the conquest of the province (with the exception of Viminacium) and all of them were located at auxiliary forts or important mining areas. We are able to suppose that these sites were definitely one of the first that received Roman newcomers[2] – army and administration. So, the first baths were probably built for people who were already used to such a facility. There is no actual way for us to confirm the presence of non-Romans at these baths. On the other hand, military and administrative staff usually depended on the local population (merchants, craftsmen, workers, etc.), so it is not completely impossible that these objects were introduced to the non-Romans for the first time in these areas specifically. In that case, it is most likely that at these very places (and some other similar ones) the non-Romans learned of the Roman way of life and its everyday routine. The earliest Roman finds could be traced to a broad vicinity of these objects. Certain hoards and graves from the Danube *Limes* indicate that some of the earliest Roman finds were integrated into local traditions. Such situations could be illustrated by the inventory of the Tekija hoard, where Roman *pateras* were located together with the silver jewelry of local craftsmen. At the same time, the handles of these *pateras* were perforated and decorated with the axe-shaped pendant of local traditions. It is uncertain for what purpose the Roman finds were transformed or used, but we can say with certainty that the whole inventory of the hoard had equal symbolic value for the owner (MANO-ZISI, 1957). This case is not unique, and similar situations were confirmed by the inventory of another Danube hoard from the same period – Bare near Viminacium (POPOVIĆ, BORIĆ-BREŠKOVIĆ, 1994). We can say that these situations reflect a change in the material culture used for maintaining identity. The Roman luxurious objects (silver) together with other local silver objects testify to the introducing of new material culture into the everyday routine. Such objects could be used either for maintaining status among the local population, or also within the new "Roman" surrounding. Although these occurrences may look unrelated, it is very important that they are derived from a relatively small area over a short period of time (the last quarter of the 1st century to the beginning of the 2nd). Most likely, this area was the one where we can certainly confirm the first archaeological evidence of changes that occurred after the Roman conquest.

The group of later baths is much larger than the previous one, and far more numerous differences between the sites can be observed. The differences and similarities between the building plans, interior decoration, presence of rooms and small finds are just some of the criteria by which we can observe these baths. On the other hand, their attachment to other structures could tell us whether they were used by the military or civilians, or maybe even by both. This time, the focus was on their social context and the role they performed in everyday life. The later baths were usually built within settlements along with other public buildings throughout the province. At this time, the first luxurious baths were built, often attached to other luxurious structures – villas or palaces. Now we can argue for social activities that differed from their basic purpose, which was bathing. As we saw before, traces of game boards, musical instruments or gardening tools support the idea of baths as centers of everyday social interaction. We are not discussing changes any more, but the way of maintaining their way of life through repeating a daily routine. This does not mean that I believe that cultural transformation was done and finished in each and every area of the province; but now we can point out the urban centers with baths as an important part of the urban landscape. Going to the baths became much more

important in this period because of all the associated activities: to see and be seen, to enjoy all the benefits that the baths were offering and to engage in all kinds of social interaction (REVELL, 2008). In other words, the baths may have served as a tool for maintaining the fluid nature of Roman identity.

Conclusions

The baths that were built during the Roman period were always visible, and archaeologists have excavated various forms of water installation from the beginning of the modern discipline in Serbia. Nevertheless, little attention has been paid to their role in everyday life. Paradoxically, on the one hand it has always been stressed that the baths were very important in the everyday life of the Romans; but on the other hand, a small group of researchers has focused on the social activities at the baths. The main objective of many published articles was a description of their architectural characteristics, or interior decoration. As a result of such researches we do not have a complete picture today. The data on the distribution of small finds inside the baths' walls may be helpful in determining the purpose of the facility, or the type of activities that the visitors were engaged in. However, I believe that the data, fragmentary as they are, could be sufficient to illustrate a number of activities that occurred within the baths. Various contexts of the baths and their connections with other structures in the vicinity were enough to indicate the potential of their social role in everyday life.

The other important thing concerns the debate of cultural transformation. We have already seen that the mere presence of the material is not enough to label something as Roman or non-Roman. It is also hard to debate whether the material culture could be the conclusive evidence that changes ever occurred. But recognition of a definite routine and its everyday repetition using a new set of material cultures could be good indicator of social change. If someone builds a bath-house in a province, that does not mean that the local population will embrace it at once and use it in the same way as did the senators in the public baths in Rome. It could be used for something completely different or in a different way. But if we are able to recognize the actual practice that was repeated in everyday routine in the same way, then we can discus social change.

If we are to put everyday life as a focus of cultural transformation research, I believe that leisure time is going to be an important part of this research. Tracing the social changes would be much easier if we tried to follow practices that people chose to engage in (board games, baths, gladiatorial combats, theaters...). Certainly, that does not mean that people became Roman by simple adoption; but they were definitely much closer to achieving and maintaining Roman identity.

Bibliography:[3]

BOJOVIĆ, D., 1975
Prilog urbanoj istoriji Beograda u periodu rimske dominacije, Godišnjak grada Beograda XXII, 5-25.

BOJOVIĆ, D., 1977
Rimske terme u parku na Studentskom trgu u Beogradu, Godišnjak grada Beograd XXIV, 5-20.

CASELLA, E. C.; FOWLER, C., 2004
The Archaeology of Plural and Changing Identities, New York: Springer.

CUNJAK, M., 1996
Terme na Orašju kod Požarevca, Viminacium 10, 105-19.

CURCHIN, L. A., 2004
The Romanization of Central Spain, New York: Routledge.

ČANAK-MEDIĆ , M., STOJKOVIĆ-PAVELKA, B., 2010
Arhitektura i prostorna struktura carske palate, in: I. Popović (ed.), Felix Romuliana- Gamzigrad, Beograd: Arheološki Institut, 49-106.

ČERŠKOV, E.,1970
Municipium D.D. kod Sočanice, Priština-Beograd: Muzej Kosova i Arheološko društvo Jugoslavije.

DE LAINE, J., 1988
'Recent Research on Roman Baths', *JRA* 1, 11-32.

DE LAINE, J. 1999
Introduction. Bathing and Society, in: J. De Laine and D. E. Johnson (eds), Proceedings of the First International Conference on Roman Baths held at Bath, England, 30 March – 4 April 1992, *JRA* Supplementary Series 37. Portsmouth and Rhode Island: Thomson Shore, 7 -16.

DIAZ-ANDREU M. et al., 2005
The Archaeology of Identity. Approaches to gender, age, status, ethnicity and religion, New York: Routledge.

DUŠANIĆ, S., 1977
Aspects of Roman Mining in Noricum, Pannonia, Dalmatia and Moesia Superior, in: H. Temporini and W.Haase (eds), *Aufstieg und Niedergang der römischen Welt II*. Berlin and New York, 52-94.

DUŠANIĆ, S., 2000
Army and Mining in Moesia Superior, in: G. Alfoeldy, B. Dobson and W. Eck (eds), *Kaiser, Heer und Gesellschaft in der Romischen Kaiserzeit*, Stuttgart, 343-63.

FAGAN, G. G., 2002
Bathing in Public in the Roman World, Michigan: University of Michigan Press.

HINGLEY, R., 2000
Roman Officers and English Gentlemen, London: Routledge.

HINGLEY, R., 2005
Globalizing Roman Culture. Unity, diversity and empire, New York: Routledge.

JANKOVIĆ, A. M., 2010
Igre na tabli sa teritorije Singidunuma i okoline. Godišnjak grada Beograda LV. Beograd, 55-68.

JONES, S., 2005
The Archaeology of Ethnicity. Constructing identities in the past and present, London: Routledge

JORDOVIĆ, Č., 1999
Rimske terme u selu Bace, in: M. Vasić и D. Marinković (eds), Prokuplje u praistoriji, antici i srednjem veku. Beograd-Prokuplje, Arheološki institut i Muzej Toplice, 197-9.

JOVANOVIĆ, A., 1975
Neki aspekti problema skupnog nalaza skulptura sa Medijane kod Niša, Starinar XXIV–XXV, 57-66.

KONDIĆ, V., ZOTOVIĆ, L. J., 1974
Viminacium-rezultati arheoloških istraživanja u 1974 godini. Arheološki pregled 16, 94-7.

KORAČEVIĆ, D., 1985
Skupi, Skoplje - teritorija rano rimskog carstva i kasne antike, Arheološki pregled 24, 73-82.

LATKOVIĆ, R., DRČA, S., JANKOVIĆ-MIHALDŽIĆ, D., 1979
Medijana. Niš, Narodni muzej.

LAZIĆ, M., 2001
Terme u Nerodimlju kod Uroševca, in: M. Lazić (ed.), VESTIGATIO VETVSTATIS Aleksandrini Cermanović-Kuzmanović, Beograd: Filozofski fakultet, 247-79.

MANO-ZISI, Đ., 1957
Nalaz iz Tekije, Beograd, Narodni muzej.

MILLET, M., 1990
Romanization of Britain: An Essay in Archaeological Interpretation, Cambridge: Cambridge University Press.

MILOŠEVIĆ, G., 1999
Crkve u podnožju Hisara u Prokuplju, in: M. Vasić, D. Marinković (eds), Prokuplje u praistoriji, antici i srednjem veku, Beograd-Prokuplje, Arheološki institut i Muzej Toplice, 161-78.

MILOVANOVIĆ, B., 2008
Izveštaj sa sistematskih arheoloških iskopavanja lokaliteta Terme-Viminacijum 2004 godine, Arheološki pregled 2-3, 51-4.

NIELSEN, I., 1990
Therme et Balnea. The Architecture and Cultural History of Roman Public Baths. Part I, Aarhus: Aarhus University Press.

PAROVIĆ-PEŠIKAN, M., 1981
Antička Ulpijana prema dosadašnjim istraživanjima, Starinar XXXII, 57-73.

PETKOVIĆ, S. et al., 2005
Roman and Medieval Necropolis in Ravna near Knjaževac, Belgrade: Archaeological Institute.

PETROVIĆ, P., 1966
Novi votivni natpisi iz južne Srbije, Starinar N.S. XV-XVI, 245-51.

PETROVIĆ, P., 1984
Porečka reka, sabirni centar za snadbevanje rimskih trupa u Đerdapu, Starinar XXXIII-XXXIV, 285-91.

PETROVIĆ, P., 1987
Brza palanka-Egeta, Đerdapske sveske III, 369-77.

PETROVIĆ, P., 1993
Naissus, in: D. Srejović (ed.), Rimski carski gradovi i palate u Srbiji, Beograd, SANU, 57-82.

PETROVIĆ, P., 1994
Medijana-Rezidencija rimskih careva, Beograd: Arheološki Institut.

PETROVIĆ, P., 1999
Niš u antičko doba (Nis in antiquity), Niš: Prosveta.

PETROVIĆ, P., JOVANOVIĆ, S., 1997
Kulturno blago knjaževačkog kraja, Beograd: Arheološki institut , Narodni muzej Knjaževac.

PETROVIĆ, P., VASIĆ, M., 1996
The Roman Frontier in Upper Moesia: Archaeological Investigations in the Iron Gate Area, in: P. Petrović (ed.), Roman Limes on the Middle and Lower Danube. Cahiers des Portes de Fer. Monographies 2, Belgrade: Archaeological Institute, 15-26.

POPOVIĆ, I., BORIĆ-BREŠKOVIĆ, B., 1994
Ostava iz Bara, Beograd: Narodni muzej.

REVELL, L., 2007
'Military Bath-houses in Britain – a Comment', *Britannia* 380, 230-36.

REVELL, L., 2008
Roman Imperialism and Local Identities, Cambridge: Cambridge University Press.

STOJKOVIĆ-PAVELKA, B., 2004
Rimska kupatila, in: G. Mitrović (ed.) Voda – smisao života, Dani kulturne baštine. Beograd: Ministarstvo kulture Republike Srbije, Društvo konzervatora Srbije, 44-7.

TOMOVIĆ, M., BULATOVIĆ, A., KAPURAN, A., 2005
Žujince – crkvište. Arheološka istraživanja E 75, 1/2004, 317-53.

VASIĆ, M., MILOŠEVIĆ, G., 2000
Mansio Idimum, Beograd: Arheološki Institut.

VASIĆ, M., 2004
Bronze Railings from Mediana, Starinar LIII-LIV, 79-109.

VELIČKOVIĆ, M., 1957
Vrela u Lisoviću. Rimsko naselje, Starinar N.S. IX-X, 377-8.

VELIČKOVIĆ, M., 1958
Prilog proučavanju rimskog rudarskog basena na Kosmaju, Zbornik radova Narodnog muzeja I, 96-118.

VUČKOVIĆ-TODOROVIĆ, D., 1969
Istraživanje limesa u SR Srbiji u okviru međuakademijskog odbora, Osiječki zbornik XII, 123-39.

YEGÜL, F., 1992
Baths and Bathing in Classical Antiquity, New York : Architectural History Foundation and Massachussets Institute of Technology.

WOOLF, G., 1998
Becoming Roman: The Origin of Provincial Civilization in Gaul, Cambridge: Cambridge University Press.

WEBSTER, J., 2001
'Creolizing the Roman Provinces', *AJA* 105, Boston: Archaeological Institute of America.

Notes:

[1] This chapter was written as a part of project 177008 (*Archaeological Culture and Identity in the Western Balkans*), funded by the Ministry of Science and Technological Development of the Republic of Serbia.
[2] The term "Roman" is used for people who had already maintained their identity as "Roman", not as an actual ethnic label.
[3] For some of the references that were published in Cyrillic script, a Latin transcription has been provided.

ROMAN LAW AND ARCHAEOLOGICAL EVIDENCE ON WATER MANAGEMENT

Sufyan Al Karaimeh

Key Words: Water management, Archaeology, Roman Law, Middle East

Introduction

Land ownership and water rights through history are fundamental not only for agrarian societies but also for urban ones. This is because water and land are vital elements for the growth of communities; consequently laws throughout the ages have been developed for land ownership and water rights. Law represents power, order and discipline. Who controls law, water and land has the power to control and protect people and their interests. Throughout history each developing society grows, resulting in the problem of population and settlement expansion. Organisation and maintenance of water systems and distribution rules were needed. Therefore, administration, documentation and law were created as a basic aspect in ancient societies to keep everything under control.

The region of this research is located in an area that had been inhabited during the Greek, Roman and Byzantine times and had flourished in the Roman period. However, there is only a limited amount of historical evidence concerning irrigation and water management from the region of the Decapolis; therefore, historical sources from the Roman period which cover the whole area where Roman law was applied will be used as a parallel example. These sources refer to the Digest of Justinian (WATSON, 1998, the Digest book XXXIX, title 3, 'Water and the action to ward off rainwater'; book XLIII, title 20, 'Daily and summer water') and other sources like the Theodosian code and novels (PHARR, 1952, book XV, 2, 'Aqueducts').

Roman laws had been compiled under the supervision of the Byzantine ruler, Emperor Justinian I. The emperor lived *c.* AD 482-565, during which the *Corpus Juris Civilis* was his significant achievement. The most famous part of the *Corpus Juris Civilis* is the Digest, the others being the Code, the Institutes, and the Novels (WATSON, 1998, dig. XXIII). His work started by directing a commission to make a collection of imperial rescripts. They accomplished the mission by writing the Code in the year AD 530. Unfortunately, the original version of this code has not survived; however, the AD 534 revised code is available.

After the code was completed in AD 530, the emperor proceeded to collate the Digest. This book was a collection of the classical Roman jurist documents from the 1st century BC to the end of the first third of the 3rd century AD (WATSON, 1998, Dig. XXIII). He ordered that the Roman juristic writings be collected and abridged in a book called the Digest. The Digest was finished and issued in the year AD 533 (WATSON, 1998, Dig. XXIII). The emperor gave his order to make a new elementary textbook for students, which is called the Institutes, and it was finalized at the same time as the Digest. The Institutes is the most systematically arranged part of the *Corpus Juris Civilis*, and is written in four books.

In order to write the Digest, Emperor Justinian instructed his *quaestor*, Tribonian, to read and extract the ancient books of Roman authority, to remove all the superfluities, obsolete rules and any matters that were already recorded in the code (WATSON, 1998, Dig. XXIV). It took sixteen compilers to write the Digest over three years, wherein each compiler specified at the head of each extract the name of the author and the book in which it appeared (WATSON, 1998, Dig. XXIV).

Most laws and Roman juridical case studies that have been selected in this paper are based on the Digest. The laws that have been selected from the Digest selection are listed below:

Justinian Ulpian Edict book 53, Neratius, Quintus Mucius, Mucius, Ofilius, Marcellus, Labeo, Ulpian Edict book 53, Ofilius, Ulpian book 53, Labeo, Paul Edict book 49, Paul Edict book 49, Paul Plautius book 15,

Javolenus, Pomponius, Quintus Mucius book 32, Ulpian Edict book 70, Pomponius Sabinus book 32, Pomponius Sabinus book 34, Julian Digest book 41, Julian Minicius book 4.

Relation between the region and Roman Law

The Decapolis region was under Roman political control (WEBER, 1990, 7-8). In addition, the region contains much Roman architectural influence. These facts give us confidence that at this time Roman law was valid and applied. It is possible that indigenous laws were also used in the Decapolis region; however, these are unknown. Gadara (present-day Umm Qays) was an important city in Roman times. This is indicated by the fact that the Roman Emperor Augustus gave Gadara and its territory to King Herod of Judaea in 30 BC. Consequently, it is known that Rome exerted political influence over the region.

The Gadarenes did not agree with King Herod ruling them, because they wanted to be joined to the Roman province of Syria; so they sent a delegation to complain and negotiate with the emperor's legate, Agrippa. Agrippa was also in charge of the Romans' water supply; and Augustus' trusted colleague was presented with the *imperium* in the east (BRUUN, 2000, 587; BUTCHER, 2003, 81), showing that there was a clear relationship between Roman authority and Decapolis, and probably there was some influence on the management of water supply in the region by Rome. The Roman architectural influence on the region can be clearly seen from the theatres, streets, baths, aqueducts and other domestic buildings. After the death of Herod in 4 BC, Gadara was joined to the Roman province of Syria (WEBER, 1990, 7-8).

It is clear that there were hydro-technology developments in the Levant region during Roman times. This is indicated by increasingly detailed water laws (WIKANDER, 2000, 651) as they were collected in late antiquity, referring to water management and landownership (see WATSON, 1998). In general, because of their engineering skills and their efficient laws, the Romans were very successful at water management. It was the Roman army engineers who played a major role by spreading such knowledge throughout the whole empire (WIKANDER, 2000, 651). In the Roman period, water use and its supply were divided between inside and outside the cities and other settlements: inside settlements between public and private dwellings, and outside settlements for agriculture (irrigation, drainage). Water quality and supply for settlements and the surrounding areas were fundamental in the Roman period; therefore they secured high quality constant water supply, which was obtained from near and far distances as appropriate (HODGE, 1992, 95-9; WIKANDER, 2000, 653).

Many archaeological studies have been undertaken on water management in the city of Rome and in other Roman cities. In addition, there are many literary sources about water administration, both from the city of Rome and from the whole of the Roman world (BRUUN, 2000, 540). These give a clear picture of how people dealt with water.

Water needs were different for people living in an urban environment or in the countryside. In the countryside, agriculture and/or luxurious living often caused a greater need for water (BRUUN, 2000, 280), while people who lived in urban places were dependent on the public water supply to their settlement. Public water means that each person has the right to water from the sea, lakes, most of the rivers and some of the springs that were not located on private land. While private water means that the owner of a private water source (such as a lake, spring or well) has the right to draw the water for his/her purposes (BRUUN, 2000, 577-8). These differences give the countryside dwellers the ability to draw water from public sources like rivers, lakes, and private sources like wells or springs. In general, farmers (countryside dwellers) used more water than the urban dwellers because of their irrigation needs. Consequently, public and private water sources could be used for irrigation, but not the sea because of its salinity.

Supporting evidence of the applicability of Roman Law in the region

Archaeological remains and historical sources support the idea that Roman law was used in the region.

In Gadara were found three different segments of water systems supplying the city with water. First, there is the large overland tunnel which delivers water from several springs to the tunnels under the acropolis, through an aqueduct at the beginning of the city from the east side. Second, there are two tunnels that run under the acropolis; the lower tunnel was finished, while the upper one was uncompleted and unused. Third, different lines spread through the city and provided the area with water (KERNER, 2004, 187). The upper uncompleted tunnel and the overland tunnel were made in the early Roman period, and the overland tunnel was used until the 6th century AD.

The Romans were well known for constructing and managing water: for example, their building of many water chambers as in Pompeii and Nimes (KERNER, 2004, 191; FAHLBUSCH, 1987, 160). Gadara also had a water chamber to distribute water through pipes and openings at different heights in three different directions and to several undefined archaeological destinations inside the city (KERNER, 2004, 191).

Some water laws were complex, as shown by the text below, where references are made to the owner of a construction and then a tomb causing damage, the site becoming religious; and then under the judge's instructions the site must be restored to its previous condition. This law is possibly meant to refer to tombs where a channel was designed in such a way that if the water overflowed from the main channel, the tomb would be protected and no damage would be caused to the tomb. Ulpian, book 53:

Labeo writes that although an action to ward off rainwater can only be taken against the owner of a construction, nonetheless, if someone has built a tomb and water from it causes damage, despite the fact that the builder has ceased to be the owner because the site has become religious, the view is still to be approved that he is liable to an action to ward off rainwater, since he was the owner at the time when the work was carried out, and that if, on a judge's instructions, he restores the site to its previous condition, no action for the violation of the tomb will be taken (WATSON, 1998, Dig. XXXIX, 3, 4).

Fig 1: Tomb with protecting channel. Photo was taken from the west side.

It is interesting that in the Umm Qays survey project in 2008, a similar archaeological situation as described in this text was recorded. A tomb measuring 180 cm by 83 cm was found, located approximately 381 cm below a main channel. Furthermore, another channel, located 40 cm above the tomb (see Fig. 1), was made to ward off running water.

The right to distributed water between the inhabitants of a settlement, especially neighbours, was important to ensure equality. In Roman law the rights specified the quantities of water for each user, and the time that it was agreed to receive it. The 2nd century AD Imperial rescript shows the equitable division between farmers (WATSON, 1998, Dig. VIII, 3, 17).

Water from a public stream for irrigation of the fields must be allocated in proportion to the areas of the estates concerned, except where an owner can legally establish a claim to a larger share. It is further decreed that damage to another owner... (WHITE, 1970, 159).

An example of an inscription comes from the community of Lamasba in south Algeria, dated *c.* AD 220.[1] Major segments of the inscription deal with the allocation of water to a number of local farmers. The inscription contains 43 names (WHITE, 1970, 157-8), each of which

acquired water from the local Aqua Claudiana (a perennial spring or an aqueduct); the water supply was divided into units of time, when the sluice was opened to feed each farmer's irrigation channel (BRUUN, 2000, 580).

Evidence on papyrus has been found in Petra, which describes a dispute over properties between two neighbours who lived in a Sadaqa settlement, a former Roman garrison city, about 25 km south-east of Petra and dated AD 544 or 574. One of the matters that the litigants disputed to arbitrators is about a spring owned by Theodoros, situated on a courtyard owned by Stephanos. The conflict is about the rights to use and conduct the water by Theodoros. The arbitrators used Roman laws.

The law concerned the 'servitude' of water, which is the custom of Roman law that allowed both parties to draw water from the spring by building water conduits through each other's properties.[2] Furthermore, Theodoros could not prevent Stephanos from eventually building anything on the courtyard where the spring was situated (KAIMIO, 2001, 719-24). This evidence demonstrates that the arbitrators used Roman law. This shows that the area was under Roman political and judicial influence and that Roman laws were valid.

Other evidence which supports the idea that Roman laws were valid in the eastern part of the Empire comes from Palmyra. A Roman tax law was inscribed both in Greek and in the dialect of Aramaic that was used in Palmyra. The text shows both the new and the old regulation taxes which were dated *c.* AD 137 (MATTHEWS, 1984, 157). The new tariff of the Palmyra version was headed '*Tax law of the exchange of Hadriana Tadmor and of the water sources of Aelius Caesar*' (MATTHEWS, 1984, 175), while the old version was headed '*Tax law of Tadmor and the water sources and salt which is the city and its borders*' (MATTHEWS, 1984, 177).

The old version shows that the text does not carry the title of Hadriana, which indicates that the new regulation had been instituted after the visit of Hadrian in AD 130. This specifies a charge for water in the new regulations, including irrigation, which was used to irrigate the oasis and the gardens. This reference is found in line number 88*: for use of the two water sources, each year 800 den* (MATTHEWS, 1984, 177).

Selected Laws

In the Roman juridical court, much attention was paid to cases related to water management. The cases were often very complicated and frustrating because some citizens fiercely disputed issues concerning water rights. For this reason new laws were instituted, like an action to control run-off rainwater, the amount of daily and summer use of water, etc. These laws were meant to give each person the right to have water, the right to share it and to protect people and properties from all kinds of damage that could be caused by water.

Damage: Deviating water, Public interests, Neighbours and Land improvement

In a period of heavy rain, damage could be caused to the fields. People or farmers who owned fields always carried out work to protect their fields from rain damage. This damage could be caused naturally or by man-made constructions, as when someone caused water to flow elsewhere than in its normal and natural course; for example, if a person constructed something which made the flow greater or faster or stronger than usual, or if by blocking the flow he caused an overflow. As an illustration, in the book of Ulpian, Edict, book 53 we see described a case of water damage and subsequent action necessary to ward off the rainwater when possible damage could be caused to a field (WATSON, 1998, Dig. XXXIX, 3, 1).

Damage by deviating water

In general, every person has the right to draw water, but in some circumstances this right can be withdrawn. In situations like guiding rainwater from one place to another, such as drainage or channelling water to be collected in a reservoir, well or dam, the right is limited and should not cause damage to a neighbour's fields. As in the juridical case of Neratius in Ulpian, Edict, book 53:

Neratius writes: Where somebody has made a construction to keep out water which normally flows onto his field from an overflowing marsh, if that marsh is increased in size by rainwater, and the said water, held back by the construction in question, damages his neighbour's field, he will be compelled to remove it by means of an action to ward off rainwater (WATSON, 1998, Dig. XXXIX, 3, 1, 2).

Changing the natural flow of a stream or a spring is not permitted, even if it is necessary on some occasion, as in a dry period. This is supported by the text in Ulpian, Edict, book 53:

The question of the source of water does not have to be raised. For even in cases where water which originated on public or consecrated property flows down onto my neighbour's land and carries out some work which causes the water to be diverted onto my property, Labeo says that the neighbour is liable to an action to ward off rainwater (WATSON, 1998, Dig. XXXIX, 3, 1, 18).

Damage and the public interest

Another case where a right may be limited is when it threatens the public interest. In these instances the authorities were involved and might even include the influence of the emperor himself. For example, in Javolenus, from Cassius, book 10, it is stated that:

If a construction that causes rainwater damage is carried out on public land, no action can be brought. But if the public land intervenes between the site of the work and that of the damage, an action will be possible. The reason for this is that nobody but the owner is liable for this action (1). Water cannot be conducted across a public road without the emperor's permission (WATSON 1998, Dig. XXXIX, 3, 18, 1).

Damage and neighbours

Many problems existed between farmers who owned adjacent fields. These conflicts could be caused by drainage if the drainage water flowed into a neighbour's field that was located at a lower level. The damage that these flows caused could prevent a farmer in a lower field from performing his work, which could cause conflict. Consequently the farmers would be provoked and a legal case would be started. A case of this type was made and cited in Ulpian, Edict, book 53, where it said that:

However, Mucius says that the construction of ditches to drain fields counts as something done for the purpose of cultivating the property, but that ditches should not be constructed for the purpose of causing water to flow in one stream since one must only improve one's field in such a way as not to reduce the quality of one's neighbour's field (WATSON 1998, Dig. XXXIX, 3, 1, 4).

Furthermore, a law was made to identify those who were responsible for any damage that was caused to a field. If the damage to a field was the result of a person or group of farmers, then they were responsible for the damage and were requested to rectify any damage to the fields. An example is cited in Paul, Edict, book 49:

Cassius says that if water flowing from or onto a jointly owned piece of land causes damage, an action can be brought either by one owner against another, or by one owner against several others individually, or by several owners individually against one owner, or by several owners against several others owners. Where one owner has brought the action and the restoration of the work and the assessment of damage have been carried out, the others' right to take action is extinguished (WATSON, 1998, Dig. XXXIX, 3, 11, 1).

Damage and land improvement

Another kind of damage can be caused by digging channels and ditches to supply water to the fields, or to remove surplus water from the fields. Some kinds of digging for the purpose of fertilising the land can also cause some problems to neighbours' fields, but this kind of digging is not liable to punishment. This is supported by the text in Ulpian, Edict, book 53:

Next, Marcellus writes that no action, not even an action for fraud, can be brought against a person who, while digging on his own land, diverts his neighbour's water supply. And of course the latter ought not to bring an action for fraud, assuming that the other person acted not with the intention of harming his neighbour, but with that of improving his own field (WATSON, 1998, Dig. XXXIX, 3, 1, 12).

Each farmer has the right to improve his land. But any work carried out should not change the natural perspective of the fields and the adjacent neighbour's fields by building a construction or closing a channel which prevents the water from flowing naturally. This is supported by text in Ulpian, Edict, book 53, 13:

Again, it must be understood that this action is available both to the owner of a higher piece of land against the owner of a lower piece, to stop the latter carrying out work to prevent naturally flowing water passing down through his own field; and to the owner of a lower piece of land against the owner of a higher piece, to stop the latter causing the water to flow other than naturally (WATSON, 1998, Dig. XXXIX, 3, 1, 13).

Rights and responsibilities: Farmers and agriculture

Communities that used irrigation did not always behave socially towards each other. In some cases, difficulties were caused for example by shortage of water in the dry season, construction of channels in a way that caused flooding to adjacent land, which caused the washing away of arable soil, and other similar problems.

Farmers and agriculture

The law specified that each farmer was responsible for the channels on his land, whether he used it for cultivation or not. With reference to Ulpian, Edict, book 53:

Mucius says that if someone can plough and sow without making water channels, he is liable should he lay out such channels, even though he may be held to have done so for the purpose of cultivating a field. But if he cannot sow without making water channels, he is not liable. Ofilius, however, says that it is legal to make water channels for the purpose of cultivating a field if they are all made to run in the same direction (WATSON, 1998, Dig. XXXIX, 3, 1, 5).

These sorts of problems deterred collaboration between farmers and caused litigation. Furthermore, cultivation that used irrigation needed a lot of water, which was critical in some places because of the scarcity of supply and insufficient distribution of water between farmers. In such situations, farmers sought any possible solutions which enabled them to get water. Therefore, the law encouraged farmers to collect rainwater on their properties, as in Ulpian, Edict, book 53:

The same authorities say that everyone has the right to retain rainwater on his own property and to channel surface rainwater from his neighbour's property onto his own, provided that no work is done on someone else's property, and that no one can be held liable on this account, since no person is forbidden to profit himself so long as he harms nobody else in so doing (WATSON, 1998, Dig. XXXIX, 3, 1, 11).

Farmers who own both cattle and land should have the right balance to grow crops and feed the animals. Animals could be fed on stubble after the harvest, but for the rest of the year they had to be stalled and fed on whatever fodder could be raised (WHITE, 1970, 152). Also, pastures did not last for the whole year. To overcome this problem one of the techniques used was to convert a piece of land into a meadow. Columella (*c.* 4 BC-AD 65) described how to restore a run-down pasture into a meadow. He suggested that after the first scything the meadow should be irrigated, but only if the soil is on the heavy side (WHITE, 1970, 26,152-3). Also in Ulpian, Edict, book 53:

Ofilius says that if someone who previously irrigated his field at a fixed time in the year makes a meadow there and starts to cause damage to his neighbour by persistent irrigation of it, he is liable neither to an action against anticipated injury nor to one to ward off rainwater, unless he has levelled the site and, as a result, the water has started to flow more quickly onto the neighbour's property (WATSON 1998, Dig. XXXIX, 3, 3, 2).

Property: Ownership, Water schedules and servitude

Properties were sometimes transferred from one person to another by inheritance or by purchase. If the property was sold inclusive of the water rights, these should not change when the sale took place. Only with agreement between the purchaser and the vendor could a change be made in supplying water, but the water rights and the ownership of the land went hand in hand and could not have different owners. In addition, any change to the rights must not harm the neighbours' land. As in Paul, Edict, book 49:

Where a property is subject to an addictio in diem, *the consent of both purchaser and vendor must be sought to ensure that the cession of water rights is made with the owner's consent, whether the property remains with the purchaser or reverts to the vendor. In a case of cession of water rights, consent is required not only from the person to whom rights over the water belong but also from the owner of the sites involved, even if the latter cannot make use of the water, since all rights over the land may revert to him* (WATSON, 1998, Dig. XXXIX, 3, 9 and 9, 2).

Ownership

An owner of a spring or channel could grant water rights. If the owner wished to grant water rights to more than one person, he might do so in two ways. Firstly, he could make an informal private agreement with the parties for the mutual use of the water; or a formal contract could be set up in law specifying the amount of water each person could draw. Julian described the right of a person to draw water from a channel, and an owner of a spring to have the right to distribute the water from one channel to two farmers. As in Julian, Digest, book 41:

I have ceded to Lucius Titius the right to draw water from my spring. The question is: May I cede to Maevius also the right to draw water along the same channel? If you think it possible to cede water along the same water channel to two persons, how should they use it? He replied: just as a right of way on foot or with cattle, or of a road, can be ceded to several people either together or separately, so the right of drawing off water can rightly be ceded. But if those to whom the water is ceded cannot agree how to use it, it will be only fair that an utile judicium *should be delivered, as it has been held should be done for dividing a common usufruct among the several people to whom it belongs* (WATSON, 1998, Dig. XLIII, 20, 4).

Some laws were needed or instituted in specific regions to share water sources in areas where there was a scarcity of water, or to keep the water in balance between farmers. Most of the cases were disputes over water rights. People claimed many rights involving water which caused the supreme authorities to create, institute or to use an old constitution to draw a clear line between the claimers and to give each person his rights.

Water schedules and servitude

As a group of farmers could benefit from one source of water, a system of water grants was established to avoid any possible dispute. The way to secure the right of having water at the time it was needed was fundamental and needed a high degree of cooperation between the members who received the water from wherever it came from. A schedule in these cases could be made and followed. The schedule was sometimes inconvenient for some farmers, for example whether the water was divided between the farmers per period, per day or per hour. If a person received water twice, once at night and once during the day, it could be difficult for him to manage to get water at both periods because perhaps other agents also had the right to the water at the same time. Furthermore, if obtaining water involved a lot of work, the farmers should decide whether to receive water during the day or night. As in the case of Paul, Plautius, book 15:

If a servitude for conducting water at night has previously been granted to me and then, later and by a separate concession, the right to conduct water by day as well is allowed to me but, during the specified period, I only use the right to conduct it at night, I forfeit the servitude for conducting water by day, because in this case there are several servitudes with different origins (WATSON, 1998, Dig. XXXIX, 3, 17).

Each person had the right to obtain water. In some cases a person did not have the right to get water, for example, if a source of water was located on a neighbour's land or it flowed over his land and the person was not the owner of the source. The person who owned the land and the water source had the right to stop the water flowing onto a neighbour's land, and he was not liable to any complaint from the neighbours. Pomponius, Quintus Mucius referred to this in book 32:

If water which has its sources on your land bursts onto my land and you cut off those sources with the result that the water ceases to reach my land, you will not be considered to have acted with force, provided that no servitude was owed to me in this connection, nor will you be liable to an interdict against force or stealth (WATSON, 1998, Dig. XXXIX, 3, 21).

If the farmers agreed on a plan to divide and distribute water between themselves and any kind of authority, then the schedule or the agreements would be seen as law and statements, and each farmer must follow the agreements, and it was not permitted to change the schedule or misuse it. As in Pomponius, Sabinus, book 32:

If I have the right to draw off water during day or night hours, I cannot do so during any other hour than the ones to which I am entitled (WATSON, 1998, Dig. XLIII, 20, 2).

It seems that the schedule might contain not only the names of the farmers but also when their turn was to draw off water, and the amount of the water that any farmers might receive. The amount of water varied daily between summer time and winter time. This was because during summer there could be a shortage of water, while in the winter there was plenty of water. In Julian, Minicius, book 4 described the different situations of using daily water between summer and winter time:

As it is settled that water may be divided not only by times but also by measures, at the same time one person may draw off daily water and another summer water on the understanding that the water is divided between them in summer, but that in winter the one with the right to daily water may draw it off by himself (WATSON, 1998, Dig. XLIII, 20, 5).

If a person drew water from a source, and did not steal, force the flow or disobey any law, then no one could prevent him from drawing this water. Ulpian, Edict, book 70:

The praetor says: Insofar as you have this year drawn off water in question not by force or stealth or precarium *from such a one, I forbid force to be used to prevent you from drawing it off in this manner* (WATSON, 1998, Dig. XLIII, 20, 1).

Purpose of water use

Two ways of using water have been noted under the juridical cases: household water and water used for irrigation. Household water was used in houses by dwellers both in urban or rural areas for drinking, washing etc. Irrigation water was used for daily cultivation of the fields. For all citizens water was available all year round, each person being allocated a fixed amount according to their circumstances. This amount could not be changed in dry periods like in summer time, even if this was difficult for farmers because of the higher temperatures which caused shortages in water supply. If a warm water supply, e.g. natural spa, was available, it was normally used for bathing; however, in certain circumstances this water could be used for irrigation. These situations could result in disputes between citizens.

One of the solutions that Labeo wrote about is distinguishing between cold and warm water. Warm water might be used for irrigation in areas that were needed. Changing the way of using water was not permitted without official authority, and the person who did that would be liable. In Ulpian, Edict, book 70:

It is asked whether only such water is included under the edict as belongs to the irrigation of fields or also what is for our use and convenience. The law we follow is that these too are included. On account of this, even if someone wishes to draw off water for town properties, this interdict may apply (12). Besides, Labeo writes that even if water is not drawn off for a farm, then because it may be drawn off to any place, the interdict still applies (13). Labeo also writes that even though the praetor in this interdict means cold water, interdicts are not to be refused for warm water, as there is a necessary use for these waters too. For sometimes they are cooled and provide a use for the irrigation of fields. Furthermore, in certain places the water is warm and is needed for irrigating the fields, as at Hierapolis; for it is a fact that the Hierapolitans in Asia irrigate their fields with warm water. And even in the case of water which is not necessary for irrigating fields, no one will doubt that these interdicts will apply (WATSON, 1998, Dig. XLIII, 20, 11, 12, 13).

The rights of exploitation (taking advantage of) and using water by any person were complicated, because things might be prohibited in some situations but allowed in others. Farmers did not always have the right to draw water for irrigation, but water could be drawn for several purposes. The owner of cattle had the right to draw water for his thirsty animals; also a person could draw water for his pleasure. As the law in Pomponius, Sabinus, book 34:

We make use of this right so that water may be drawn off not only for irrigation but for herd animals or as an amenity (WATSON, 1998, Dig. XLIII, 20, 3).

Indigenous laws

The potential existence of indigenous laws concerning irrigation and water management in the region of this research is possible, as these laws existed before the Roman influence on the area. These can be seen in the Roman law of Emperor Arcadius and Honorius Augustuses to Asterius, count of the Orient. The law encouraged indigenous people to preserve their ancient water rights and not disturb them through any innovation or change.

(15.2.7) Emperor Arcadius and Honorius Augustuses to Asterius, count of the Orient. (After other matters.) We decree that ancient water rights that are established by

long ownership shall remain the property of the several citizens and not be disturbed by any innovation. Thus each man shall obtain the amount that he has received by ancient right and by custom lasting to the present day. The punishment shall remain which was provided for persons who wrongfully used secret channels of water for the irrigation of their fields or for the beautification of their gardens. Given on the kalends of November in the year of the consulship of Caesarius and Atticus November I, 397 (PHARR, 1952, 431).

Conclusion

Roman laws have been demonstrated in this study to explain the picture of water management in the city of Gadara, one of the Decapolis cities, and its territory. The region is quite similar to the region of Italy on which the laws were based: it is hilly and has a lot of springs. These laws have been chosen as a parallel to studying the Gadara region from historical sources. Although the historical sources are limited for this region, historical and archaeological evidence has been gathered from several locations to support the subject of this study. Roman water rights were elaborated and developed because of the flourishing of Roman legal thought (BRUUN, 2000, 557).

The Roman law of Iulius Frontinus, which was instituted in AD 100, gave private grants in the city Rome to the rest of Italy and wherever Roman law was applied (BRUUN, 2000, 579). These grants would be included into the laws of the Roman Empire that expanded around the whole of the Mediterranean region, including the Decapolis region. Furthermore, Roman political and architectural influence on the region and the archaeological and historical evidence from the region and from the whole empire support the idea that Roman law might have been used in cases that had to do with irrigation and water management in the Gadara region.

Water problems are almost universal. Most cases show the common problems of how to get, distribute and allocate water supplies between people. The case of the commune of Lamasba in south Algeria mentioned above is a clear example of this. The process of conveying water from the source to its final destination can sometimes cause damage to others. People disputed the way to deal with water, and it could end in conflict. Therefore, the law was used to solve water problems.

The law could be indigenous or imperial; in this study the imperial Roman law was used. The lack of historical sources about issues relating to water management in the Gadara region does not mean that water problems did not exist. The code of the Emperor Arcadius and Honorius Augustuses to Asterius, count of Orient demonstrates that indigenous people might have had their laws and methods to deal with water issues. These indigenous laws could have been similar to Roman laws.

The region of Gadara is the same as other places dealing with water problems. The archaeological evidence shows that the city of Gadara needed water in both Hellenistic and Roman times. An underground tunnel system was built and hydrological knowledge augmented. Irrigation systems were made on the slopes of the surrounding mountains. These facts demonstrate that the city itself and its territory expanded, showing a possible growth in the population and economy. These developments may have caused the society some problems, and water management was likely to be one of them.

Roman law presented cases that could have happened in any society dealing with water. As the Gadara region was under Roman political control, it was sensible to use Roman laws to solve issues concerning water. At the same time Roman law protected ancient water rights and gave a new vision to dealing with cases which were complicated and frustrating, as in the case of the papyrus from Petra mentioned above.

This study gives evidence from several regions of the entire Roman Empire, which shows some communities using Roman law in cases related to water problems. This indicates that the situation was probably valid for the city of Gadara. It is unknown whether the society had its own laws, but the possibility that Roman law was used is highly probable. Finally, laws concerning water rights, water management and agriculture are meant to prevent and protect any person or properties from damage or harm caused by man-made activities, and to stop any illegal intervention. Consequently, people were encouraged to apply the law.

Bibliography:

BRUUN, C., 2000
Water legislation in the ancient world (c. 2200 BC - c. AD 500), in Ö. Wikander (ed.), *Handbook of Ancient Water Technology*, Leiden: Brill, 539-607.

BUTCHER, K., 2003
Roman Syria and the Near East, London: British Museum Press.

FAHLBUSCH, H., 1987
Elemente griechischer und römischer Wasserversorgungsanlagen, in: *Die Wasserversorgung antiker Städte: Geschichte der Wasserversorgung* Bd. 2, Mainz, 133-221.

HODGE, A. T., 1992
Roman Aqueducts and Water Supply, London: Duckworth.

KAIMIO, M., 2001
Petra inv. 83. A settlement of dispute, in: I. Andorlini, M. Manfredi, G. Bastianini and G. Menci (a cura di), Atti del XXII Congresso Internazionale di Papirologia 2, Firenze, 719-24.

KERNER, S., 2004
The water systems in Gadara and other Decapolis cities of Northern Jordan, in: H. D. Bienert and J. Häser (eds), *Men of Dikes and Channels, the Archaeology of Water in the Middle East*, Rahden: Verlag Marie Leidorf GmbH, 187-202.

MATTHEWS, J. F., 1984
'The tax law of Palmyra: evidence for economic history in a city of the Roman East', *JRS* 74, 157-180.

PHARR, C., 1952
The Theodosian Code and Novels and the Sirmondian Constitutions. A Translation with Commentary, Glossary and Bibliography, New York: Greenwood Press.

WATSON, A., 1998
The Digest of Justinian, Philadelphia: University of Pennsylvania Press.

WEBER, T., 1990
Umm Qays, Gadara of the Decapolis, Amman: Al Kutba Jordan Guiders.

WHITE, K. D., 1970
Roman Farming, London: Thames and Hudson.

WIKANDER, Ö., 2000
Historical context. The socio-economic background and effects. The Roman Empire, in: Ö. Wikander (ed.), *Handbook of Ancient Water Technology*, Leiden: Brill, 649-61.

Notes:

[1] WHITE, 1970, dated the inscription to AD 100, while BRUUN, 2000, dated it to AD 220.
[2] A servitude consisted in the right of a person, other than the owner, ... to make a certain use of another's land. From BRUUN, 2000, 582.

WATER DISTRIBUTION IN PELUSIUM - A SHORT NOTE ON A LARGER PROBLEM

Krzysztof Jakubiak
Institute of Archaeology
University of Warsaw

Key Words: Pelusium, Tell Farama, water cisterns, sewage systems, bathhouse, baptistry

The city of Pelusium (the modern Tell Farama), located at the mouth of the Pelusiac branch of the Nile, doubtlessly belonged to the most important seashore cities in the eastern Mediterranean. The site, very significant from a historical point of view, is unfortunately still very little known from its archaeological remains. So far the several archaeological expeditions which conducted the fieldworks there were not able to excavate the most important buildings of the large city that Pelusium originally was. The purpose of the text below is an attempt to show how the water distribution system was used by the citizens of Pelusium. The present knowledge has given us an opportunity to take a closer look at that matter.

Pelusium, a city that developed along an island, was, according to Herodotus, founded by pharaoh Psamtik I who settled people here from Caria and Jonia. Since that time the city constantly developed and flourished until the Arab conquest of Egypt. After that event Pelusium was slowly abandoned.

Since the excavation strategy mainly focused on unearthing the most important monuments, any elements associated with water distribution and its usage by the Pelusium citizens were not the main aim of the excavators. However, several constructions were unearthed during the researches that were strictly associated with water usage and distribution in the city area. The truth is that no human society and no city are able to function without fresh water sources. Consequently, even the smallest settlement needed to have at least a simple but effective system of water administration. If for the smallest settlement water was one of the essential supplies for its functioning, then the same was even truer for much bigger municipal structures. Pelusium is a good example of a city whose functioning was disorganizedly associated with water.

The basic and most essential question is how the Pelusians replenished their water for daily life and for economic purposes. The answer to that problem seems to be very simple, considering the fact that the city was flourishing on an island in the middle of the Pelusiac branch of the Nile. In other words, fresh water was drawn and distributed directly from the river, and more sophisticated systems such as aqueducts were not necessary in Pelusium. However, the distribution of water in the city itself was a separate problem. Let us focus on the area west from the so-called Great Theatre in Pelusium, where a Polish-Egyptian expedition was recently conducting excavations. There, in the centre of the ancient city, was located a complex of large cisterns collecting huge amounts of water needed for Pelusium's functioning.

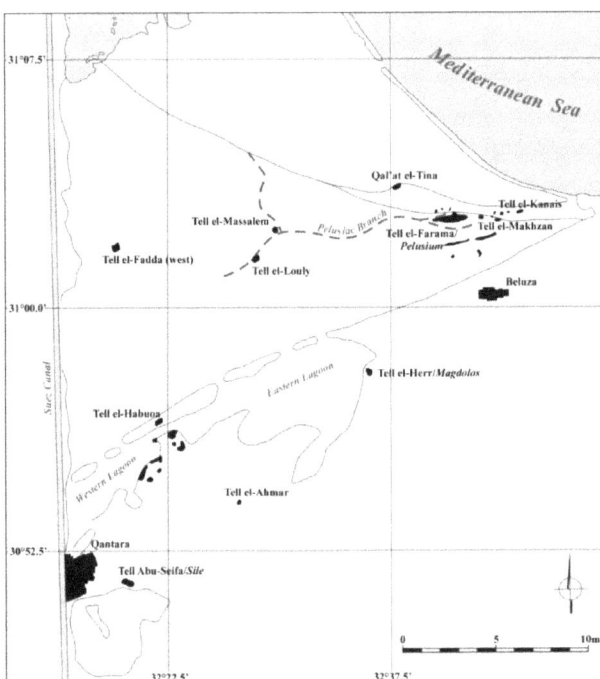

Fig. 1 Pelusium: Site locality

Fig. 2 Pelusium, Tell Farama: the water cistern in the central part of the city

characteristic and monumental in the Tell Farama landscape. The cistern was originally rectangular in layout, which most probably covered the area of at least two *insulae*. In recent years the structure was partially investigated by a Swiss expedition (DELAHAYE, 2005, 299-305). Thanks to their fieldwork results, it is possible to draw a preliminary plan of that important building. Further research will bring a better understanding of the whole layout of the cistern complex and more precise dating for the structure. According to present knowledge, this hydrological construction was functioning during the 4th and 5th centuries C.E. We should remember, however, that the pottery evidence discovered during preliminary excavations dated the latest phase of the construction back to the 2nd century C.E. If these traces are accurate, the cistern building may be associated with the earlier urban planning development that took place over that time. Most probably this centrally situated cistern distributed water for several important public buildings functioning in the city. The public structures which certainly needed a lot of water were constructed nearby. One of those buildings was erected to the north-west of the cisterns: the so-called northern bathhouse, which was functioning there (ABD EL-MAQSOUD, 1984-5, 3-8).

This municipal construction undoubtedly played a key role in the water distribution system. That is why the monumental structure was built not only in the centre of the city, but also in the centre of the island, and not without reason. Unfortunately the eastern part of the complex is badly damaged, but still the ruins are the most

Fig. 3 Pelusium, Tell Farama: the so-called Northern Bathhouse, a view from the south-east

Fig. 4 Pelusium, Tell Farama: the so-called Southern Bathhouse, a view from the north-east

This building was constructed between the Late Roman fortress and the banks of the Nile, where it can be assumed that landing piers were built. The bathhouse situated here most probably used water drawn directly from the Nile. The distance to the river was never farther than 50-60 metres. A debatable question is how the water was drawn from the river. Supposedly a *saqieh* system was in use here, as the most effective and popular in Ancient Egypt. There may also have been a special well functioning next to the bathhouse.

On the opposite side of the city, south of the Late Roman fortress, a much larger bathhouse was partially unearthed. According to the excavators this thermal complex was called the Southern Bathhouse. The building is characterized by a relatively complicated structure (BONNET, CARREZ-MARATRAY, ABD EL-SAMIE et al., 2006, 371-84; BONNET, CARREZ-MARATRAY, ABD EL-SAMIE, et al., 2007, 247-60.) Besides elements typical for baths to function, such as water pipes, stoves, hot air canals and basins, numerous chambers used for different purposes were discovered.

From the point of view of urban planning, the place was originally occupied by a *gimnasion*. When this disappeared in the city landscape, the space was used for the bathhouse construction (CARREZ-MARATRAY, 2006, 385-9.)[1]

The baths, constructed most probably in the 3rd or at the beginning of the 4th century C.E., certainly used large amounts of water daily. That is why this building that was so important for public life was erected close to the southern limits of the ancient city, just near the river course which overflowed the whole of Pelusium. Here also the *saqieh* method of drawing water directly from the Nile could have been used. The excavations are far from final answers and no such construction has been discovered so far. Also essential for the functioning not only of the bath complexes but also for ordinary structures in the whole city fabric was the Pelusium sewage system. Of course the baths were essential foci for the citizens who took special care of their daily hygiene and social life, but removing the dirt and effluents were also important tasks for the city.

Fragments of that kind of structure were excavated by the Polish-Egyptian expedition working in the area of the Great Theatre in Pelusium (A. AL-TABA'I, M. ABDAL-MAQSOUD, P. GROSSMANN, 2003, 271-83; GAWLIKOWSKI, 2004, 67-72; JAKUBIAK, 2003, 2004, 73-5; JAKUBIAK, 2005, 61-8; MAŚLAK, 2005, 69-71; JAKUBIAK, 2006, 125-35).

Fig.5 Pelusium, Tell Farama: the so-called Southern Bathhouse, the western part of the building

Directly east of the theatre building, part of a Hellenistic sewage system was unearthed. This, probably the oldest sewerage discovered till now in Pelusium, was constructed of red brick bounded in clay mortar. The sewage canal was built on a foundation of brown clay, identified as a Hellenistic rubbish dump. Inside the canal some pottery fragments were found which supported the proposal of dating it to the Hellenistic period. The most significant fact is that the system was out of order, or its functioning was over by the time the theatre was constructed. This monumental structure almost totally cut off and blocked the older sewerage. In other words, the older Hellenistic city infrastructure was almost completely devastated and forgotten. The building remains which were unearthed east of the theatre, just near the sewage system, were most probably also from Hellenistic times. Among the damaged remnants it was possible to recognize that this monumental building was undoubtedly connected with the above-mentioned main drain. The building had its own canalization system, and the dirty water was channelled away from the building directly to the sewerage. The size of the building and the hydrological solution used here indicate that this construction could have been of public use and probably one of many others in this part of the ancient city. The buildings were erected along the flagstone paved street which had its own drainage system transporting liquid waste, most probably directly to the Pelusiac branch of the Nile.

Fig.6 Pelusium, Tell Farama: the sewage canal situated east of the Great Theatre

To the north of the northern façade of the theatre, a fragment of another drainage system was discovered (JAKUBIAK, 2006, 128-9). Here the sewage system was constructed of red brick and bounded in clay and lime mortars. The technique of construction was different in this case.

- 2,11

- 1,47

- 1,86

- 1,75

- 1,18

- 1,21

- 1,05

- 1,53

- 2,25

- 1,59

- 1,60

- 1,48

- 2,14

- 1,50

- 1,89

- 1,33

P

P

- 1,87

- 1,22

- 1,30

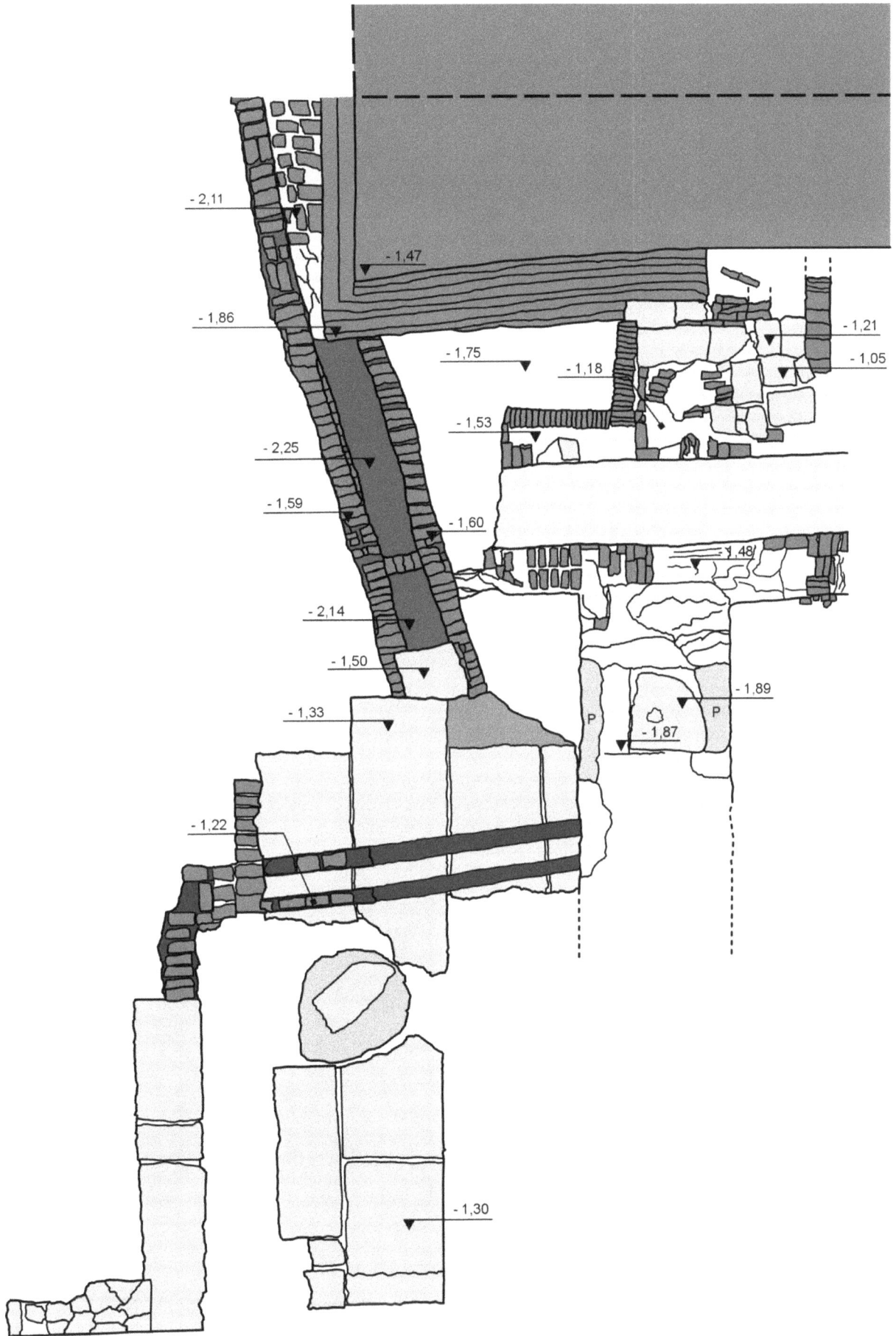

Fig. 7 Pelusium, Tell Farama: sewage canal, east of the Great Theatre

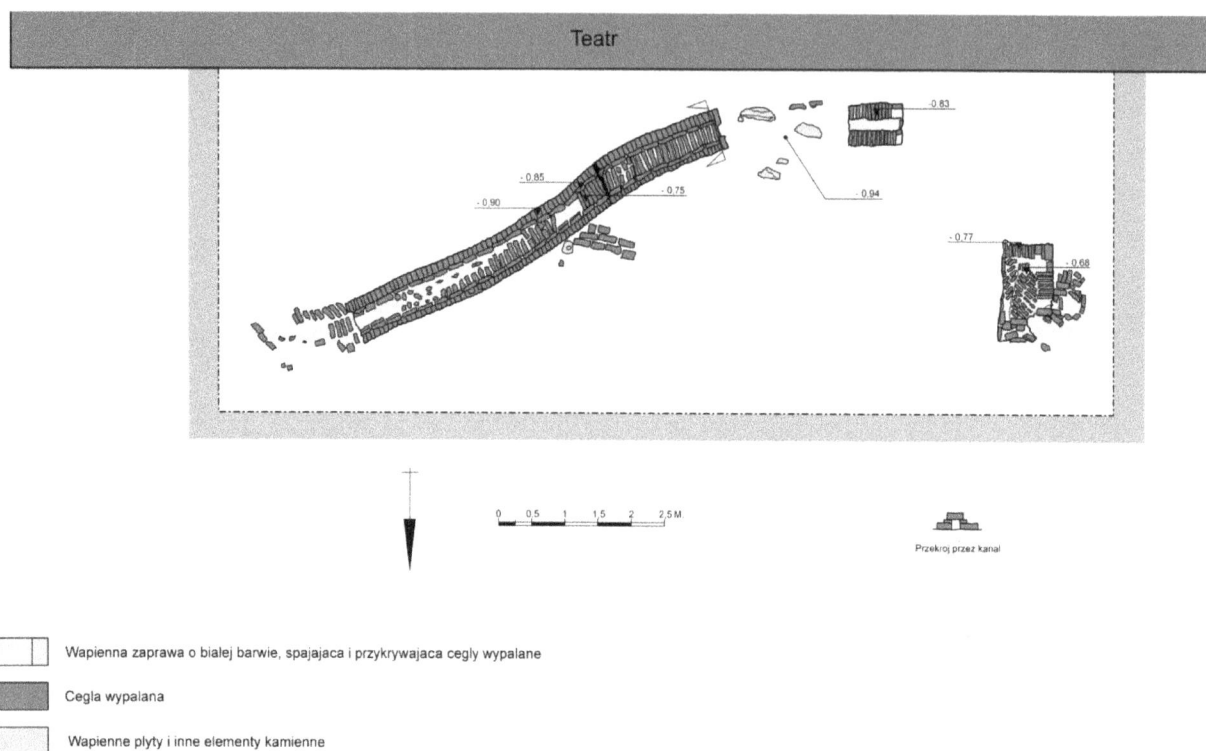

Wapienna zaprawa o bialej barwie, spajajaca i przykrywajaca cegly wypalane

Cegla wypalana

Wapienne plyty i inne elementy kamienne

Fig.8 Pelusium, Tell Farama: A sewage canal north of the Great Theatre

The drain was much smaller and covered not with lime flagstones but with red bricks. The canal did not run along the straight line but formed a structure similar to the shape of an S. The whole construction was placed under the street which was running in an east-west direction along the theatre façade. Originally the street was covered with flagstones. In some places traces of eroded lime flagstones were recognized on the surface. Red bricks covering the drain remained until the present day in the city pavements, in the form of red lines. Red, clearly visible bricks had an additional function: they made the drain easy to recognize and easy to open when cleaning was necessary. This system was functioning during the Roman period and corresponds to the building of the theatre. This partly excavated and fragmentarily recognized drain was without doubt a small section of a much larger city sewage system constructed anew in Pelusium during the Roman period. The changes were most probably involved by rebuilding the *insulae* arrangement. At the same time, the older Hellenistic infrastructure was gradually replaced as not being effective enough for the flourishing city of the 2nd or 3rd centuries C.E.

The last evidence of a water distribution system in the theatre area was excavated inside the building. After the theatre had been abandoned, a supposedly quick devastation process started, as we can assume observing the unearthed pulpitium area (GAWLIKOWSKI, 2004, 67-72; K. JAKUBIAK, 2004, 73-5; JAKUBIAK, 2005, 61-8).

The devastation was very systematic and affected the whole theatre building. The material from this monumental structure was reused in other Late Roman buildings constructed in Pelusium. The theatre was certainly furnished with many stone elements, like flagstones, wall decorations, seats in the auditorium and many others.

Fig. 9. Pelusium, Tell Farama: the Great Theatre, the technical canal in the *pulpitium* area

Some of them were probably incorporated in new structures. However, we should remember that marble and limestone could have been very easily melted into lime in special kilns. For this process water was also necessary. A canal discovered in the area of the *pulpitium*

54

was probably constructed for that purpose. This suggestion can be supported by the character of the construction: the drain was mainly built from reused material. Even some Hellenistic architectonic decorative elements were used for covering the drain. Moreover the pottery discovered during the excavations supports our suspicions that the devastation process was intensified in the 5th and 6th centuries C.E.

In the central part of Pelusium, south-east from the theatre, a partially excavated area can be observed; another sewage system was unearthed there by the Egyptian archaeologists. Besides the canals, the so-called southern street with remains of dwelling architecture was discovered there (JAKUBIAK, 2010, 65-74). Unfortunately the results have remained unpublished, so that little can be said about the buildings apart from the fact that they were attached to the sewage system. Moreover, a so-called 'pipeline' is connected to a relatively large red brick building, on the upper floor of which a water collector was originally functioning. If the observations are correct, the structures provided water for the buildings in that part of the city. The method for water distribution used here was based on gravity – the natural pressure of water lifted to the first floor. The system let water flow under relatively high pressure at a long distance from the 'water tower'. Most probably the whole construction was built up in the Late Roman period.

Other fragments of a water canal or a water pipeline were discovered in the western part of the city behind the hippodrome (ABD AL-MAQSOUD, TABA'I, GROSSMANN, 2001, 17-20). By contrast to the systems mentioned above, serving the dwellings and public areas of Pelusium, the canal discovered here functioned in the industrial zone of the ancient city. The hippodrome which was located on the city outskirts was constructed in Late Antiquity, probably in the 4th or 5th century CE, and the industrial area of Pelusium needs to be dated to the same period. The water for manufacturing purposes was drawn directly from the river, but the exact method of drawing it is unfortunately not known. Excavations in this part of the city were conducted on a limited scale, so our knowledge about the industrial zone is rather small.

Most probably one of the products manufactured there was *garum*, a type of fermented fish sauce. Water was a crucial element in the *garum* production process. In the bordering area, traces of metallurgical workshops were also discovered.

Fig. 10 Pelusium, Tell Farama: the Great Theatre, a view of the industrial canals

Fig. 11 Pelusium, Tell Maqsan: the Baptistry, a view from the west

In the western part of Pelusium is also located the so-called 'church with an atrium'. Here in the ruins situated west of the Late Roman fortress a very large baptistry was brought to light (EL-TAHER, GROSSMANN, 1997, 255-62; GROSSMANN, HAFIZ, 2001, 109-16.). The baptistry had been built up with a cross-shaped layout and placed in the north-western part of the church. Water is obviously necessary to the Christian baptism ceremony, so the system of providing water was a subject of great importance in the church. Yet there are no traces of canals or pipelines surviving in the building. Most probably water was transported to the church directly from the Nile, possibly by donkeys or other animals. This would have been easy since the river course was situated no farther than 100–120 metres north of the church.

Fig. 12 Pelusium, Tell Maqsan: water cistern with a *saqieh* construction

In the eastern part of Pelusium, on the Tell Maqsan, another monumental building with (so far) the best preserved water system was excavated. It is probably the largest, or second largest church complex in the city. A huge baptistry was discovered inside (ABD EL-SAMIE, CARREZ-MARATREY, 1998, 127-32; BONNET, ABD EL-SAMIE, 2000, 67-96; BONNET,

ABD EL-SAMIE, 2003, 75-93; BONNET, ABD EL-SAMIE et al., 2005, 281-91). The construction was situated in the south-eastern part of the sacral complex. It is rectangular in layout with an apse on the narrower side placed in the eastern part of the construction. An entrance for the catechumens was situated in the southern elevation of the church complex and led directly to the baptistry. This baptistry is one of the biggest ever discovered in northern Egypt. Consequently a huge quantity of water was necessary to fill it up. Besides the baptistry there were additional constructions discovered nearby that required large amounts of water: the convent and the latrine installation located near the main entrance to the sacred complex. This is supposedly the reason why, a few dozen metres to the west of the church, a big cistern was constructed (BONNET, ABD EL-SAMIE, et al., 2005, 281-91). The cistern was able to hold water not only for the church complex's daily functioning but also served the pilgrims who frequently visited this place.

It is worth noting that the cistern is the only construction of this kind fully excavated in the whole of Pelusium. Fieldwork confirmed that the water transported to the cisterns was drawn from a well or from the Nile, using a very effective lifted water *saqieh* system. This ancient method of water distribution is well known in the Near East, even in present times.

The examples presented above show that our knowledge of water distribution and sewage systems is very fragmentary and disordered. To date no correlated project focusing on that very important problem of how the city functioned has been conducted. However, the results of a few years of fieldwork led by several international and Egyptian expeditions are quite interesting, considering the fact that the excavations were conducted only in several parts of this large city. Some important conclusions appeared on the subject of the Pelusium sewage system. First, it seems very possible that all the water used in the city was drawn from the Nile. That observation is quite obvious, since Pelusium was developed on an island. Moreover, other sources of fresh water were probably too far to connect them with the city by an aqueduct system. The water from the Nile was drawn and used directly from the river for technical and industrial purposes only. In other cases, when water was destined for consumption (everyday use by the Pelusium citizens) or for baths, it was necessary to collect it in large water tanks.

At least two large cistern buildings were recognized in the urban space of Pelusium. These structures not only collected fresh water for any kind of use, but what seems to be the most important reason for their construction was that the water inside got cleaned in the most natural way. Water taken directly from the Nile was muddy and dull, but when it was left in the cistern all the dirt, mud and other heavy sediment accumulated at the bottom of the basin. After several days or weeks the water was clear, cool and ready to use. The cisterns' location in the middle of the city could also be very helpful in case of fire.

Fig. 13 Pelusium: the water cisterns in Tell Maqsan

So far we do not know whether or how the cisterns were connected to the important buildings in Pelusium, such as the bathhouses or the church from Tell Maqsan. We can only speculate that, considering the functions of the above mentioned buildings, such a system could exist. Unfortunately no clear evidence can be given right now to support this supposition.

The other question is how the sewage canals functioned in the city itself. The only two fragments of sewage systems discovered in Pelusium date from the Hellenistic and Roman periods. Since both fragments were unearthed in the central part of the city, we can assume that the sewers were constructed as a municipal investment and covered a large area of Pelusium. The canals were constructed under the network of streets, which proves that they were designed as part of the overall urban planning. Thanks to the observations of the stratigraphical positioning and correlations between both investments, we can also notice that the oldest city sewage system had become ineffective in Roman times when Pelusium was larger and flourishing. Rebuilding the drainage canal network probably happened in parallel to the major rearrangement of the city centre. The theatre, some new streets and divisions of the *insulae* were erected in this part of the city, together with a new drainage system. The Roman canals network was in use until the central part of the city had been abandoned, which is dated to the time shortly before Chosroes II's invasion of Egypt in 619 CE. The deserted houses were destroyed and probably used as a source for building materials for the Late Roman camp construction which had been erected to the west of the theatre remains (ABD EL-MAQSOUD, 1984-5, 3-8; ABD EL-MAQSOUD, CARREZ-MARATRAY, 1988, 97-103; ABD EL-MAQSOUD, EL-TABA'I, GROSSMANN, 1994, 95-103; ABD AL-MAQSOUD, TABA'I, GROSSMANN, 2001, 17-20).

To summarize: the main city sewage system in Pelusium was functioning from Hellenistic times until the beginning of the 7th century CE. Of course, the system did not remain unchanged during that time. It must have been restored, cleaned and enlarged several times. The proper functioning of the canal network was very important for Pelusium, a city located on muddy and boggy terrain. The drains served to remove not only liquid waste but also extra water which penetrated the lowermost parts of buildings. However, the only trace of canal rearrangement that can be observed in archeological layers is the one mentioned just above.

To conclude: the water management system in Pelusium can be divided into three groups. The first is the installations providing water for industrial and technical purposes. These types of constructions were unearthed in the western part of the city and inside the abandoned theatre area. The second category is the municipal investments – all the hydraulic and hydrotechnical structures, such as city canals and pipelines. As was mentioned above, we don't know much about these constructions.

The last category of water city infrastructure is the large water cisterns which supplied fresh water to the citizens of Pelusium for their everyday needs and for the public baths. The Late Roman or Early Byzantine water cistern situated near the church in Tell Maqsan served as a freshwater distributor for this Christian cultural complex.

Bibliography:

ABD EL-MAQSOUD, M., 1984-5
Preliminary Report on the Excavations at Tell Farama (Pelusium), First two seasons (1983/4 and 1984/5), ASAE 7 (1984-5), 3-8

ABD AL-MAQSOUD, M., TABA'I, A., GROSSMANN, P., 2001
New Discoveries in Pelusium (Tell Al-Faramā), a Preliminary Report, BSAC XL, 2001, 11-20

ABD EL-SAMIE, M., CARREZ-MARATREY, J.-Y., 1998
L'église de Tell El-Makhzan à Péluse, in: D.Valbelle, Ch. Bonnet (eds), 'Le Sinai durant L'Antiquité et le Moyen Âge, 4 000 ans d'histoire pour un désert', Actes du Colloque «Sinai» qui s'est tenu à l'UNESCO du 19 au 21 Septembre 1997, Paris 1998, 127-32

AL-TABA'I, A., ABD AL-MAQSOUD M., GROSSMANN , P., 2003
The Great Theatre of Pelusium, in: N. Grimal, A. Kamel, C. May-Sheikholeslami (eds), 'Hommages à Fayza Haikal', IFAO, Bibliothèque et d'Étude 138, 2003, 271-83

BONNET, CH., ABD EL-SAMIE, M., 2000
Les églises de Tell el-Makhzan, Les campagnes de fouille de 1998 et 1999 (avec des annexes de Ch. Simon, P. Ballet et V. Bardel, J.-Y. Carrez Maratrey), CRIPEL 21, 2000, 67-96

BONNET, CH., ABD EL-SAMIE, M., 2003
Les églises de Tell el-Makhzan, La campagne de fouille de 2001 (avec une annexe de D. Dixneuf), CRIPEL 23, 2003, 75-93

BONNET, CH., ABD EL-SAMIE, M. et al., 2005
L'Ensemble Martyrial de Tell Makhsan en Égypte, GENAVA, Revue d'Histoire de L'Art et d'Archéologie, Musée d'Art et d'Histoire, LIII, 2005, 281-91

BONNET, CH., CARREZ-MARATRAY, J.-Y., ABD EL-SAMIE, M. et al., 2006
L'Église Tétraconque et les Faubourgs Romains de Farama à Péluse (Égypte – Nord Sinai), GENAVA, Revue d'Histoire de L'Art et d'Archéologie, Musée d'Art et d'Histoire, LIV, 2006, 371-84

CARREZ-MARATRAY, J.-Y., 2006
Une Inscription Grecque à Dédicace du Gymnase de Péluse, GENAVA, Revue d'Histoire de L'Art et d'Archéologie, Musée d'Art et d'Histoire, LIV, 2006, 385-9

DELAHAYE, F., 2005
Un Complexe de Cisternes à Péluse (Égypte – Nord Sinaï), GENAVA, Revue d'Histoire de L'Art et d'Archéologie, Musée d'Art et d'Histoire, LIII, 2005, 299-305

EL-TAHER, R., GROSSMANN, P., 1997
Excavations of the Circular Church at Faramā-West, MDAIK 53, 1997, 255-62

GAWLIKOWSKI, M., 2004
Tell Farama (Pelusium), Preliminary Report on a Season of Polish-Egyptian Excavations, PAM 15, Reports 2003, 2004, 67-72

GROSSMANN, P., HAFIZ, M., 2001
Results of the 1997 Excavations in the North-West Church of Pelusium (Faramā-West), BSAC XL, 2001, 109-16

JAKUBIAK, K., 2004
Preliminary Remarks on the Stratigraphy and Pottery, PAM 15, Reports 2003, 2004, 73-5

JAKUBIAK, K., 2005
Tell Farama (Pelusium), Preliminary Report on the Second Season of Polish-Egyptian Excavations, PAM 16, Reports 2004, 2005, 61-8

JAKUBIAK, K., 2006
Tell Farama (Pelusium), Report on the Third and Fourth Seasons, PAM 17, Reports 2005, 2006, 125-35

JAKUBIAK, K., 2010
Tell Farama, Pelusium. City Urban Planning Reconstruction in the Light of the Last Researches in: J. Popielska-Grzybowska, J. Iwaszczuk (eds), Proceedings of the Fifth Central European Conference of Egyptologists. Egypt 2009: Perspectives of Research. Pułtusk 22-24 June 2009, Pułtusk 2010, 65-74

ŁAJTAR, A., 2007
A Fragment of an Opisthographic Slab from Tell Farama (Pelusium), Palamedes, A Journal of Ancient History 2, 2007, 203-6

MAŚLAK, S., 2005
Some Terracotta Figurines from Tell Farama (Pelusium),
PAM 16, Reports 2004, 2005, 69-71

VALBELLE, D., BONNET, CH. (eds), 2010
*Le Sinai durant L'Antiquité et le Moyen Âge, 4 000 ans
d'histoire pour un désert*, Actes du Colloque «Sinai» qui
s'est tenu à l'UNESCO du 19 au 21 Septembre 1997,
Paris 1998

Notes:

[1] Moreover, our attention should also be directed to another inscription
found in the northern part of Pelusium which mentioned the partially
preserved names of Pelusium's gymnasiarches. For details, see
ŁAJTAR, 2007, 203-6.

WATER SUPPLY IN PALMYRA, A CHRONOLOGICAL APPROACH

K. Juchniewicz, M. Żuchowska
Institute of Archaeology
University of Warsaw

Key words: Palmyra, Syria, water supply, archaeology

Introduction

It is a quite well known fact, that Palmyra emerged in the middle of the Syrian Desert because of availability of water. Water system of the city consisted of springs, aqueducts, water towers, pipelines and many other features, had never been a subject of separate monograph. The first who challenged the problem of putting together data from different sources was Crouch (CROUCH, 1975, 151-186). In her article she presented history of research as well as described all features connected with water that she had surveyed during her visit to Palmyra. Contribution of Crouch is still valid although some new research had been made. Articles of Meyza (MEYZA, 1985, 27-33) and Barański (BARAŃSKI, 1997, 7-17) shed new light to chronology and nature of Western Aqueduct and pipelines in the Camp of Diocletian. Test trench in the Great Colonnade made by Żuchowska in 2002 gave us information about stratigraphy and chronology of water lines under the street. Excavations conducted by Schmidt-Colinet in the Hellenistic Quarter revealed new data concerning water management in Hellenistic period (SCHMIDT-COLINET, AL-AS'AD, AL-AS'AD,2008, SCHMIDT-COLINET, 2010).

In this paper authors would like to present state of research on water system of Palmyra in chronological order as well as to introduce results of their survey in 2010.

Geological structure

Before we start to describe the water management it's worth to mention some facts about availability of water in this region. This problem was researched and published only once by a geologist G. Carle in 20' of the past century. According to his research there were two types of water accessible in the city: the first – "thermal water" emerging at least in two places – source Efqa and source called "Serai", characterized by the relatively high temperature of 29 Celsius degree, and the second – "superficial water" characterized by lower temperature, about 22 – 23 degree,, protected by the layer of the crystallized limestone rock and supplied by the rainwater (CARLE, 1923, 153 -154). It seems that Carle considered sources lying in the vicinity of the city and supplying Palmyra with the fresh water as the "superficial water". It's impossible now to determine if he was right, or not, because during the last decades all of them dried out completely. Generally, the geological structure of terrain around Palmyra is characterized by the impermeable, but full of fissures and grottoes layer of crystallized, limestone rock which let water penetrate on the long distances and help the rain water to survive underground on some level.

Early Period

It seems obvious that the abundance of water was a main condition that caused people to settle in the vicinity of Palmyra. The source of sulphuric water lying on the south-western edge of the Palmyra's area , Efqa seems to be in use already in the Neolithic period – the traces of Neolithic occupation was found near the entrance to the underground grotto called Ain al-Efqa (DU MESNIL DU BUISSON, 1966, 162). In the bronze age, however, the settlement concentrated on the opposite – south-eastern edge of the ruins zone, under the vestiges of the roman temple of Bel (DU MESNIL DU BUISSON, 1966, 181 – 184).

There had to be an alternative source of water, used by the population of the bronze age village. It seems possible that it was the second sulphuric source mentioned by Carle and called "Serail source" lying north-west of the hill where this settlement developed. Crouch discuss the identification of the source by Carle mentioning her talk with J. Starcky who assured the source really existed, but the water was sweet, not sulphuric. Crouch suppose they meant two different sources (CROUCH, 1975, 157). However it was water was abundant in the vicinity of the bronze age settlement.

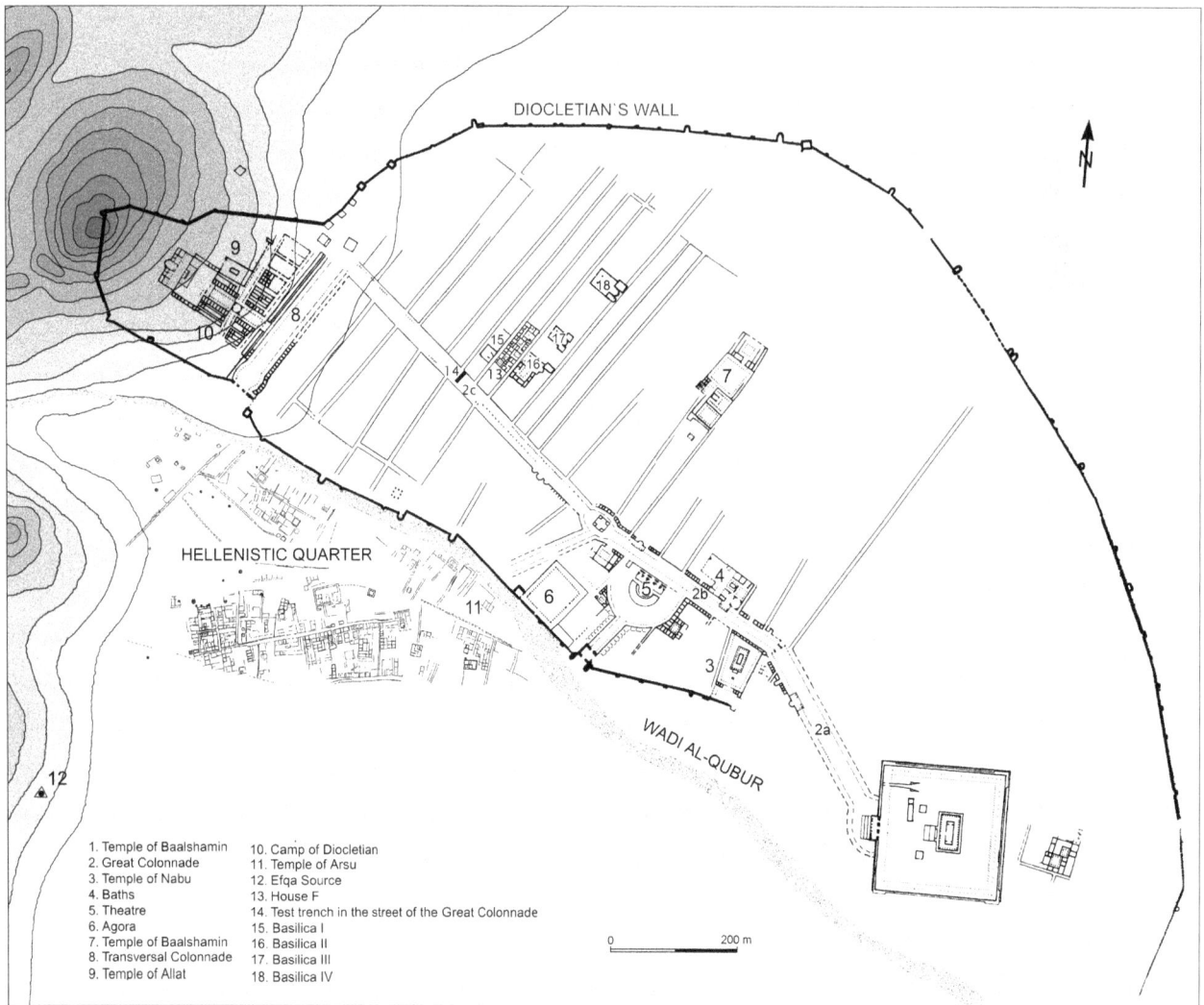

1. Temple of Baalshamin
2. Great Colonnade
3. Temple of Nabu
4. Baths
5. Theatre
6. Agora
7. Temple of Baalshamin
8. Transversal Colonnade
9. Temple of Allat
10. Camp of Diocletian
11. Temple of Arsu
12. Efqa Source
13. House F
14. Test trench in the street of the Great Colonnade
15. Basilica I
16. Basilica II
17. Basilica III
18. Basilica IV

0   200 m

Fig.1 Plan of Palmyra

Hellenistic Period

Even though the presence of the vestiges of the Hellenistic quarter of Palmyra was supposed by the scientists and hypothetically located in the area south of the Wadi al-Qubur since 70' of the last century (GAWLIKOWSKI,1973, 9), only the last two decades of research, archaeological prospection and fieldwork, gave us any evidence of Hellenistic architecture in the city.[1] Thanks to the geophysical prospection (AL AS'AD; SCHMIDT – COLINET, 2000, 61 – 93 pl. 7-16; PLATTNER; SCHMIDT-COLINET, 2010, 417-427) as well as aerial photography interpretation (DENTZER; SAUPIN,1996, 297-318) we are able to follow the urbanistic shape of this part of the city, but restricted area of the archaeological excavation in this zone makes any research on the water supply very difficult. It's definitely too early to build any general view of the water management in the city in the Hellenistic period, we can make however few observations which are worth to note and which can give us a basis to the further research on the later systems.

The only water – related structures which can be connected with the Hellenistic layers of occupation are

the well, and a pipeline, found in the sounding trench effectuated by the Syro – German mission in the middle part of the so-called *Hellenistic quarter*. The construction of the well dates back to the II c. B.C. According to the preliminary publication, it was well preserved, almost 2 m. wide and 15 m. deep structure, carefully lined with the limestone blocks (SCHMIDT-COLINET; AL-AS'AD KH.; AL-AS'AD W. 2008, 455 – 459;Abb. 3,4,6. PLATTNER; SCHMIDT-COLINET, 2010, 418-420, Abb. 3-5.,7-9). It was found in the middle of one of the main streets, which implies an existence of an earlier urban layout of which the well made a part. As it was written in the previous chapter, the geological structure of terrain enabled the possibility of use of the sulphuric water present in the layer of the limestone rock. The porous texture of the rock let the water penetrate far from the source as long as the source existed. But more probably the water in the well was the fresh superficial water supplied by the rain. At the bottom of the well in the Hellenistic quarter, water was found in the time of excavation, even if the sulphuric source dried completely few years earlier. Consequently, wells were the popular supplementary source of water in later, better known periods, abundant inside of houses, as well as in the public space. The structure found in the Hellenistic

62

quarter however is much bigger, deeper and better constructed than the later examples what implies that it was rather a primary than a supplementary water source for the Palmyrean population in the II c. B.C.

Two pipelines observed in the sounding trenches are dated by the excavators on the turn of the II cent. B.C. and Augustean period respectively (SCHMIDT-COLINET; AL-AS'AD KH.; AL-AS'AD W. 2008, 455 – 459; Abb. 3,4,6. PLATTNER; SCHMIDT-COLINET, 2010, 418-420 Abb. 3-5.,7-9).[2] Both are superposed over the opening of the earlier well and follow the trace of the main street, oriented more or less East – West. It is impossible to assume which source supplied the aqueduct of which the pipelines consist small parts. At present the only possible conclusion is chronological: at the first stage of settlement in Palmyra the main source of water supply consisted a well (or more probably – the wells), but already at the end of the II or at the beginning of the I c. B.C. citizens of Palmyra found a source of fresh water ant built aqueduct distributing water to the Hellenistic quarter of the city.

Roman Period

The vestiges of the Roman period in Palmyra are very abundant and much better examined, there are however still many blank spaces in our knowledge about water management in the city.

According to the *Palmyrean Tariff,* the tax law of Palmyra, there were two sources of water supplying the city in that period. In Aramaic version of the text we read: [LTŠ]MYŠ 'YNN TRTN DY M[Y] DY BMDYT' D 800 – "For use of two sources of water in the city – 800 denarii." (CIS II, 3913) Unless the lively discussion between the scientists we still don't dispose of any deciding evidence which could let us consider which sources were those, mentioned in the inscription. Actually, we have evidence of existence at least few sources of water in the proximity of the city: source Efqa, described above, unknown source of water in the proximity of Bel Temple, used by the population of the bronze – age village, source at Gebel Rueisat, few kilometers West of Palmyra, the source of Biyar al-Amye, north of the city, and a source between Palmyra and the village Sedat on the south-west.

Source Efqa in roman period seems have rather other than economic function. No traces of installation intended for water distribution was found in the neighboring of the source entrance. Du Mesnil du Buisson suggested that the cult of Yarhibol was performed nearby (DU MESNIL DU BUISSON, 1966,), and the hypothesis seems to be confirmed by some epigraphical data (GAWLIKOWSKI 1974, no 125, 127; GAWLIKOWSKI, 1973, 115-116). Moreover, many altars consecrated to the anonymous god were found in the proximity of the entrance to the grotto, where source flow out to the surface, so the place had to be considered as holly (GAWLIKOWSKI 1974, no 106 – 127). It could also be used as public bath, as it was still used 30 years ago, before it dried out. The monumental

staircase leading down to the grotto and some architectural decoration preserved *in situ* seam to confirm both hypothesis. The water was of course draw up from the water layer in the Efqa grottoes and natural underground corridors through the shafts or wells, but according to the present state of knowledge, no aqueduct guiding water to the city was built in roman period.

Fig.2 Ancient entrance to the Efqa grotto

Source at Gebel Rueisat is the only well known and the relics of aqueduct connecting the source and the city are partially well preserved. The source lays 9 kilometers West of the city and is not visible now, but it was described by Carle in 20th of the last century. According to his description the water was cought at the foot of the hill by five canals carved in the rock and flowed down to the underground pool. Then it was conducted to the city through the underwater aqueduct (CARLE, 1923, 155-156). The aqueduct itself was described already by Wood in XVIII century (WOOD, DAWKINS, 1753) and then researched by Meyza in 80' (MEYZA, 1985) and Barański in 90'(BARAŃSKI, 1997). The terrain prospection was effectuated in 2007 and 2010 by the authors. All collected data suggest that the aqueduct close to the source ran underground and was very carefully worked, with a stone channel in the middle, stone walls and stone slabs with ventilation – openings closing the aqueduct on the top. There was at least one staircase leading down to the underground corridor, documented by Wood (WOOD, DAWKINS 1753, pl. XXVII), and Barański (BARAŃSKI, 1997, 14).

Fig.3 Underground corridor of the Western Aqueduct

Closer to the city, aqueduct appear on the surface in the form of the stone channel running on the ground on the slope rising north of the wadi al Qubur and forming the northern part of the so called Valley of Tombs, or on the bridges, crossing the beds of the perpendicular wadis. This part of the aqueduct seems to be relatively late and it's elements date to the period between end of IIIth till the VIIIth century. In the middle of the wadi however Barański found the terracotta pipeline which he supposed to be an alternative, maybe earlier line of the western aqueduct (BARAŃSKI, 1997, 14). Without large scale excavation it is impossible to determine whether the water supply system connected with the source in Rueisat was one, very complex and many times rebuilt structure or we find fragments of two (or even more) separate structures. The detailed description of all survived parts

of aqueduct is published by Barański. The aqueduct had to be used in the roman times – the relics of the architectural decoration and statue found on the Gebel Rueisat were dated by Barański to the II c. A.D., and he suggested this date as approximate moment of construction of the aqueduct (BARAŃSKI, 1997, 15).

The third source is the Biyar al-Amye (now called by the local people Umm al Biyara) north of the city. The place, where the source flow out is presently incorporated by the modern military base so it's impossible to make any research on it. The localization was however described by Carle (CARLE, 1923, 156). We do not dispose of any evidence of use this source in the roman period.

Finally, the most mysterious south – western source. Tourtechot in XVIII century, described a place somewhere south – west of Palmyra, where people stopped to water the camels Few steps were leading down to the underground aqueduct made of stones, marked with the inscribed Greek letters (TOURTECHOT 1735, 341). The place described by Tourtechot could be an entrance to the underground aqueduct analogical to this, described by Wood on the West. No trace of this aqueduct survived till now, but it seems well possible that it could supply the southern – Hellenistic quarter of the city and was still in use in the Roman times. The suggestion of Crouch that the aqueduct had to be Hellenistic or Justinian because of use of the Greek letters is rather misleading (CROUCH, 1975, 166), since the Latin alphabet was used very rarely in Palmyra, even in the Roman period while Greek was quite common. Definitely the area south west to Palmyra was abundant in water. Carle mention few villages 10 – 20 kilometers from Palmyra where he noted the sources of water which he calls "superficial water" (CARLE, 1923, 156).

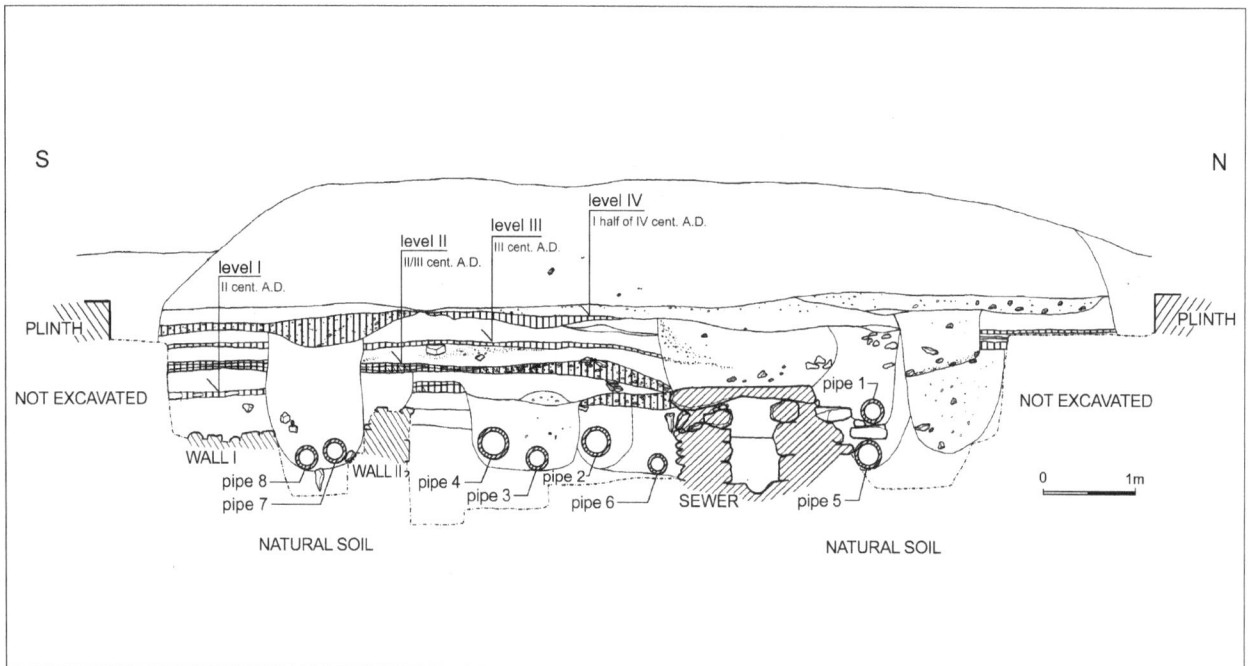

Fig.4 Cross- section of the test trench in the street of the Western section of the Great Colonnade

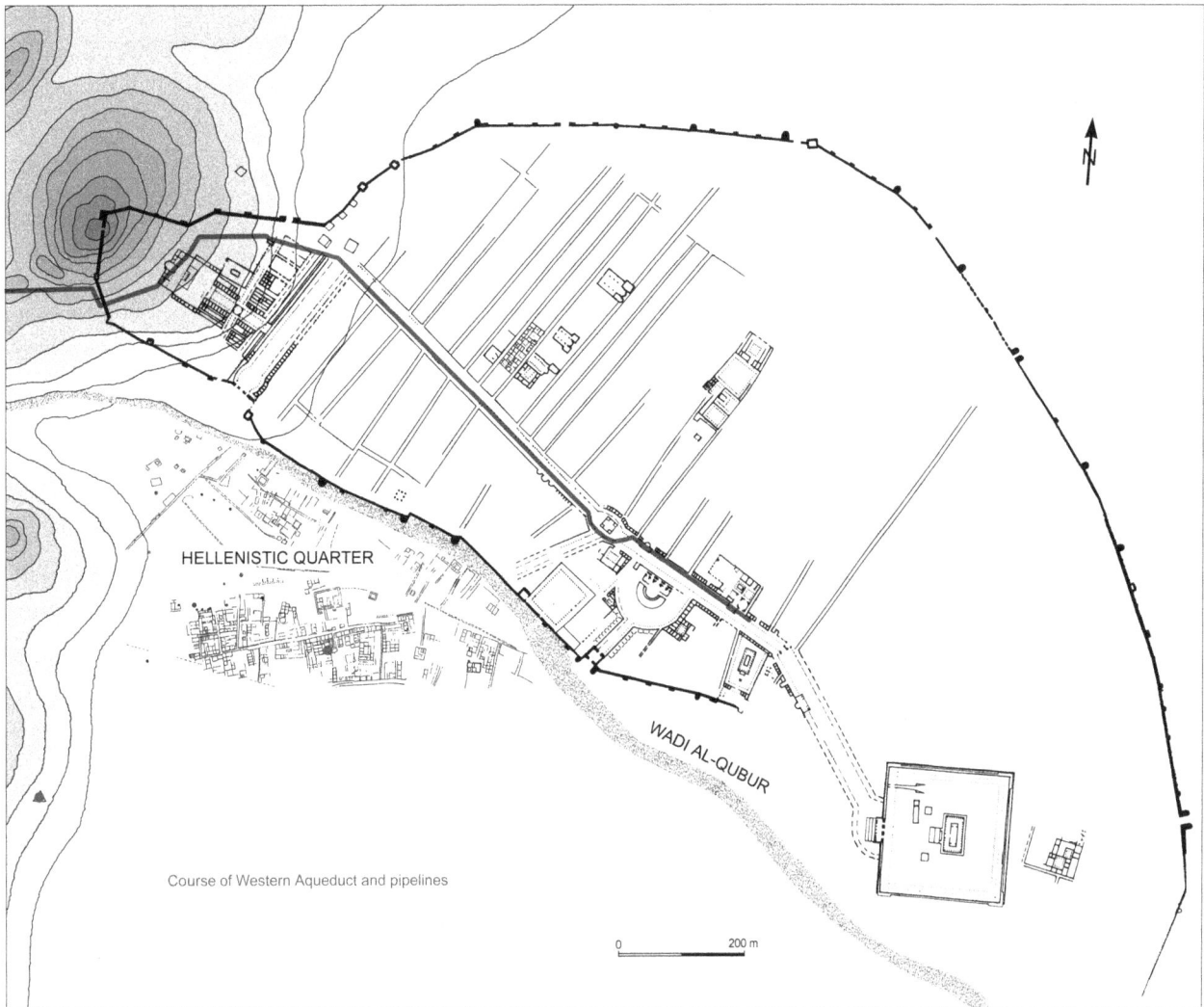

HELLENISTIC QUARTER

WADI AL-QUBUR

Course of Western Aqueduct and pipelines

0   200 m

Fig.5 Water–related structures dated on the Roman period

There was at least one more source, located in the Camp of Diocletian, not very abundant, it seems however that it started to be in use when the camp was build.

There is no way to decide which two of these sources were mentioned by the text of the *Tariff*. If we interpret literally the phrase "sources in the city" (CIS II, 3913, 58), it should mean the sources *intra muros*. From this point of view only source Efqa and the source in the vicinity of the Bel Temple, that had to be in use since the bronze age – possibly the one mentioned by Carle as the *source of Serail* can be discussed. *Sources in the city* could mean also places on the territory controlled by Palmyra, like this at Rueisat. The text of *Tariff* mention the amount of 800 denari for "administration" or simply "use"of the two sources. Teixidor believed this was a yearly price for an access to the water for irrigation (TEIXIDOR, 1984, 74), the same opinion was presented also by Matthews (MATTHEWS, 1984, 177). Teixidor proposed to determine source Efqa and source at Rueisat as two mentioned by the text of the Tariff, in our opinion, however, these two had in the ancient times completely different range and functions. Up to now, there is no evidence of use of Efqa source itself for any economical or agricultural purposes in Roman times, but it was used

for watering gardens in the later times, even in the XXth century (CROUCH, 153) Source at Rueisat supplied the city with fresh water, which was rather too precious to use for gardens. In our opinion the lack of evidence for the other sources mentioned above makes any interpretation of the, rather vague, text of the *Tariff* incomplete, ant the question remains open for the further research.

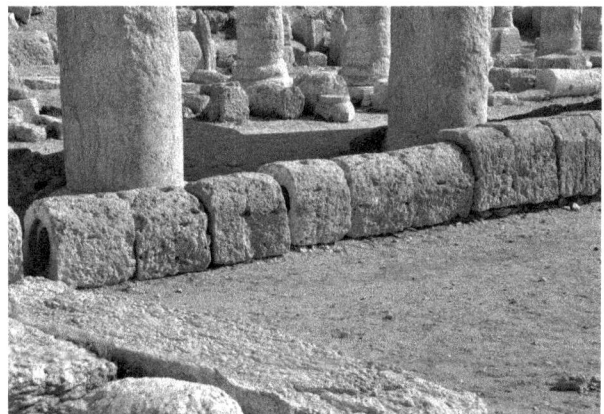

Fig.6 Stone pipeline near the Theatre, view from the West

Fig.7 *Castellum* of the Northern Aqueduct, view from the South

All we know about distribution of water in the Roman period is also connected with the Western aqueduct and the source at Rueisat. Unfortunately we do not know the place where water from the aqueduct was collected and divided – a kind of castellum had to be present somewhere on the northern border of the city, but was never found. After entering the city water was distributed through the pipelines following the line of the Great Colonnade – the main road of the city in this period. Small sounding trench done by the Polish mission in the western part of the Great Colonnade allowed to understand the chronology of this system. The earliest pipelines date to the early II c. A.D. what correspond with the date given by Barański for the construction of the Western Aqueduct. The ventilation openings in the underground aqueduct had to let some sand penetrate to the water. Consequently, pipes were often blocked and destroyed so the whole distribution system had to be repeatedly repaired and replaced by the new one. We can observe at least four phases of renovation corresponding with new pipelines and usually followed by lying new layer of the gravel on the street   The last reparation destroyed the surface of the street dated on the I half of the IV c. A.D.(ŻUCHOWSKA 2003, p. 291 – 294). It seems that system of distribution was very close to this known from Pompeii. Water was lifted to the top of water towers, and probably partly collected there creating pressure for another section of the pipeline. Water towers were almost the same type as in Pompeii (HODGE, 1996, 271, FIG. 12.). Stone bases of piers probably built of bricks and measuring about 1.2x1.1m, with two vertical

pipes are still visible along the northern portico of the Great Colonnade near the Baths of Diocletian.

Late Roman and Byzantine Period

Until now, the only fragment of Palmyra where separate archeological research of water management had been undertaken is Camp of Diocletian, and even though it is still insufficient. According to Meyza, who examined few sections of Western Aqueduct inside the Camp, the aqueduct was definitely constructed during the reign of Diocletian (MEYZA 1985, 32-33). As we mentioned above, this chronology was questioned by Barański, but the Late Roman chronology for the section of the which passes through the Camp seems to be correct. Barański established typology of pipes, based on the results of the Polish excavations (BARAŃSKI 1997, 8-13). According to his research the majority of pipelines in the Camp are dated between III c A.D. and a half of V c. A.D. (BARAŃSKI 1997, 8-9). Some of them had to be connected with the main line of the Western Aqueduct, which ran along the slope of Jebel Husayniyet and under the *Principia* building (DASZEWSKI, KOŁŁĄTAJ, 1977, 74-77). They ran through so called Water Gate, located in the eastern corner of the Camp, carrying water for the rest of the city (BARAŃSKI, 1997, 9-12).

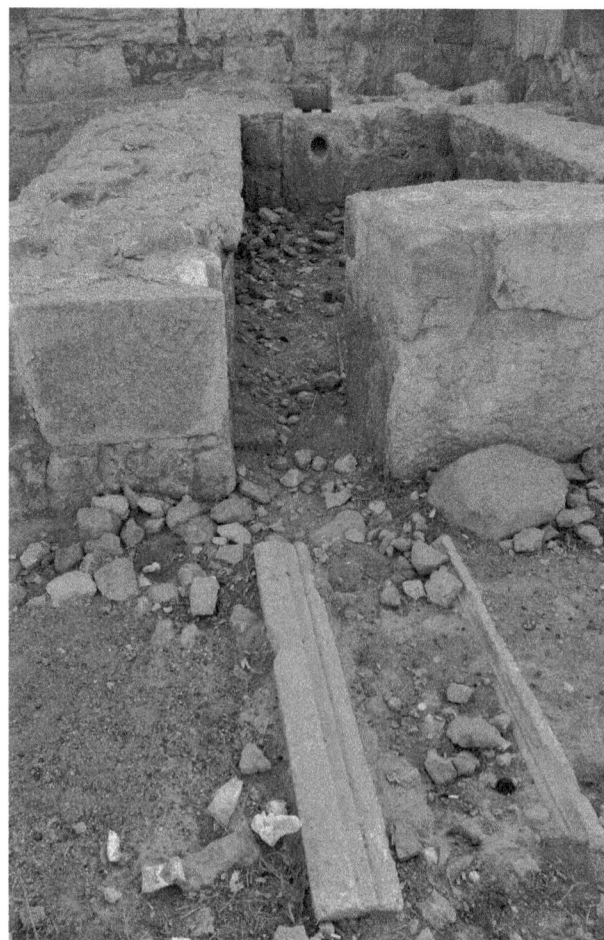

Fig.8 *Castellum* of the Northern Aqueduct, view from the North

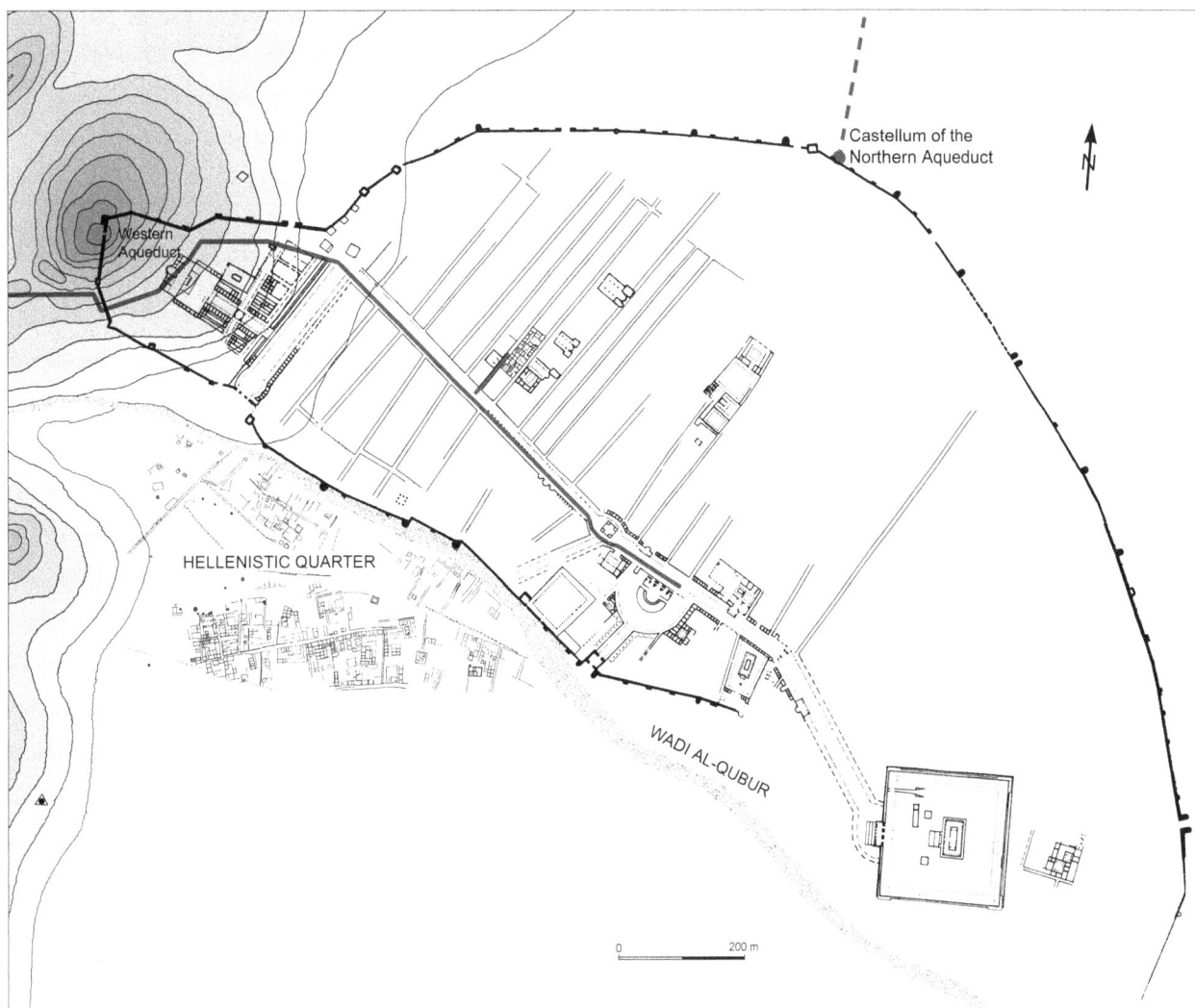

Fig.9 Water–related structures dated on the Late Roman and Byzantine periods

Small spring, mentioned above, was found between *Principia* and northern line of fortifications. It was catched into square stone structure, partly subterranean. Water was collected in a cistern about 1.3m deep, accessible by steps and door in the eastern wall. According to Barański, this structure was separate and does not seem to have been linked to any known waterline (BARAŃSKI, 1997, 9). It is very probable, that this source of water were used for the needs of the Camp only.

In the Late Roman and Byzantine period distribution of water from the Western aqueduct generally followed the earlier system. As it was said before two new water pipes were added probably in the middle of the IV c. A.D. (ŻUCHOWSKA 2003, 291 – 294) Afterwards the new system of water distribution, made of stone pipeline running on the street level was built. It is usually attributed to the building activity of Justinian in this region, described by Procopius (PROCOPIUS, *De aedificiis*, II, xi, 10-12). There is no archaeological prove of this hypothesis, nevertheless it seems to be plausible. The simple comparison of levels shows that the system is definitely later than the terracotta pipelines described above. From the other side blocks from dismantled stone

water conduit can be observed in the buildings of the Ommayad Suq, so the construction of this peculiar waterline have to be placed in the period between IV and VIII c. A.D. The manifestation of the emperor's good will would be a good occasion for creation of such strange and imposing structure. The chronology of pipes, established by Barański, seems to support such idea.

Fig.10 Stone base of water tower in the northern portico of the Great Colonnade, near the Theatre

Fig.11 Northern Channel, view from the South

The research carried by the Polish mission in 1997 in the "street of Diogenes" brought some evidence for the alternative water supply system, not connected with the Western Aqueduct, being in use in the Late Roman period. The two pipelines with a slight north – south inclination were found on the Diogenes street in the V-VI century context (ŻUCHOWSKA, 2000, fig.2, 189). It seems plausible that the pipelines were used for distribution of water from the source at Biyar al-Amye. The use northern aqueduct is now confirmed by following discoveries.

In the northern section of the fortifications of Palmyra one can find a structure which is probably a *castellum* of the northern aqueduct. Located by the tower Tt13, it is attached to the defensive wall from its outside. As well as the wall itself, the structure is partly constructed of reused blocs of limestone, originally from the tombs. The whole structure is more or less rectangular in plan. The inlet was located from the north, close to north-eastern corner. Part of an aqueduct is still visible there. It was an underground channel made of reused elements, filled with rusty-colored sand, feature which can suggest the quality of water brought by this aqueduct (JUCHNIEWICZ *et al.* 2010, 56).

The *castellum* consists of two parts. First is a big chamber where water from the channel was collected. It was barrel vaulted, as inclination of the preserved chamber walls suggests. The chamber was probably the first part of filtration system, before the water was channeled further. It is filled with dirt and trashes now so the floor is not accessible. The system of five small chambers forms the second part where the water was further filtered. Taking to consideration analogy from Pompeii and Nîmes and description from the book of Vitruvius, it is possible that the water was divided into separate flows by this system (VITRUVIUS 8.6.1-2; CALLEBAT 1973, 149-156; HAUK, NOVAK 1988, 393-407; HODGE 1992, 281-282). We must remember however that the Pompeian feature was not a model copied throughout whole Roman world, but construction which was designed to meet local needs, as Hodge stressed 9 HODGE 1996, 275). In our considerations we must also bare in mind, that Palmyren example is at least 200 years (if not more) later than Pompeian one. State of preservation of Palmyren *castellum* doesn't allow us to describe its exact functions. As far as we can observe it should be considered first of all as purification plant and device to stop water pressure to damage the pipes distributing water in the city.

Fig.12 Water tank near the Theatre, probable *Sabil*, view from the South-East

Its not certain when *castellum* had been constructed. As far as there is no archeological data we can rely only on ground observations which in fact can be quite conclusive. The *castellum* is definitely not earlier than defensive wall, which was built during Aurelian reign or shortly after (JUCHNIEWICZ). It is adjacent to the wall but not bonded with it structurally, so, in our opinion, it wasn't part of aurelianic building program. There were strong building activity during Tetrarchs, when Sossianus Hierocles was in charge of strengthening Palmyra as garrison city (SEYRIG, 1934, 7; MILLAR, 1993, 182; KOWALSKI, 1997, 44). However, the character of building technique applied in the construction of *castellum* is different than structures attributed to IV c. Another emperor, whose building activity is well known is Justinian. His works in Palmyra were described by Procopius (PROCOPIUS, *De aedeficiis*, II, xi, 10-12). Information provided by him is not entirely correct, as recent studies on fortifications has proved, but there is archaeological data according to which Palmyra in VI c. experienced considerable prosperity[3]. Thus, it is highly possible, that *castellum* has been built during that time.

Not much left of Northern Aqueduct which can be visible on the ground. Its existence is proved by *castellum* described above. Its beginning, so called *caput aque* is

located probably approximately 10 km north of Palmyra. About 7 km in this direction, in a place Biyar al-Amye there is fragment of vaulted underground channel where some Palmyrene graffiti were found (CROUCH, 1975, 166). This place is described as a source of sweet water which gives more than the spring in Diocletian's Camp (STARCKY, 1952, 82).

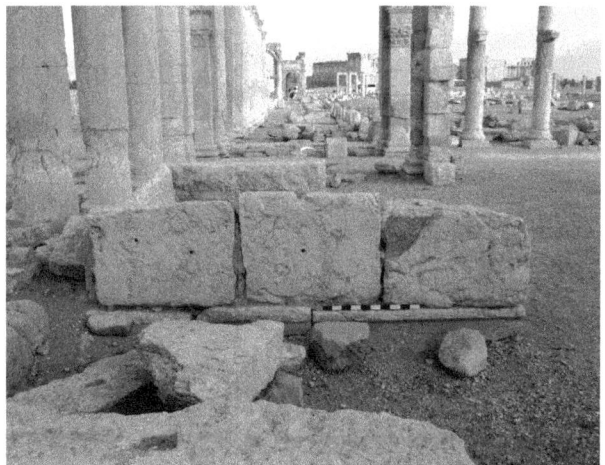

Fig.13 Water tank near the Theatre, probable *Sabil*, view from the West

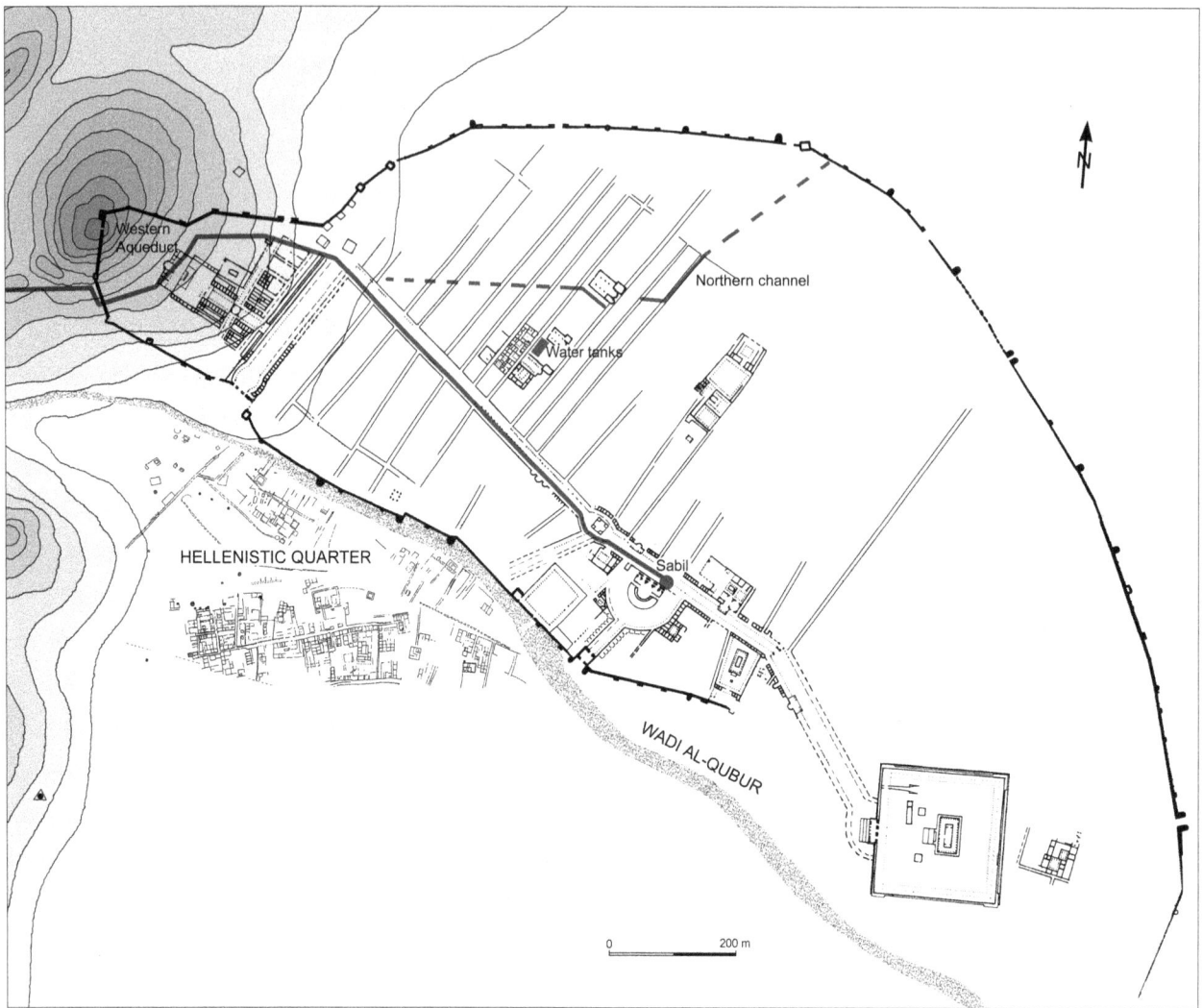

Fig. 14 Water–related structures dated on the Early Arabic Period

Structures reported by Crouch are somehow uncertain. We know them only from the interviews or talks (Du Mesnil du Buisson claimed that he discovered the place where channel enters the city, Collart described in the interview how he walked in the channel for miles) – none of this is published (CROUCH, 1975, 166). Because of establishing military base by Syrian army near Biyar al-Amye our survey was very limited and we traced probably only fragment of foggara reported by Carle (CARLE, 1923, 156). However, its not necessarily connected with the ancient aqueduct.

It is thus certain that Palmyra had its northern aqueduct. It had to be operational during Late Roman era as well as Byzantine because it was connected with *castellum*. Palmyrene inscriptions reported inside the channel suggests earlier date of construction – official inscriptions in Palmyrene language disappear in the second half of III c. A.D., probably due to change of character of the city and its administration (YOUNG, 2001, 183), so graffiti in the channel is probably not later. At this stage of research however, we can't describe exact nature and chronology of this aqueduct, as well as the system of distribution of water delivered by it. Data is very scarce and definitely insufficient.

Early Arabic Period

It seems that in Early Arabic period, which in this article we understand as period between VII and IX c. A.D., pattern of water management again followed earlier ideas. System of creating pressure inside the pipelines was the same as the one in Roman Period, and used water towers as secondary *castella*. It seems that water towers in Islamic period were located along the southern portico of the Great Colonnade, where late conduits were laid. Water channeled through those pipelines was delivered most probably by Western Aqueduct. Its section between the source at Rueisat and Abu al-Fawares was still operational until middle of the XX c.

Excavation in the Camp of Diocletian, in the area of Water Gate revealed pipeline which is dated by Barański to the Abbasid Period (BARAŃSKI, 1997, 9-10). This pipeline, as well as the earlier one dated to Omayyad Period, is clearly visible in the Great Colonnade, running along the Omayyad suq (AS'AD, STĘPNIOWSKI, 1989, 209-210). The Omayyad pipeline was replaced by the later one probably after earthquake. Some of the monumental architraves from the Great Colonnade fell down and destroyed the Omayyad conduits.

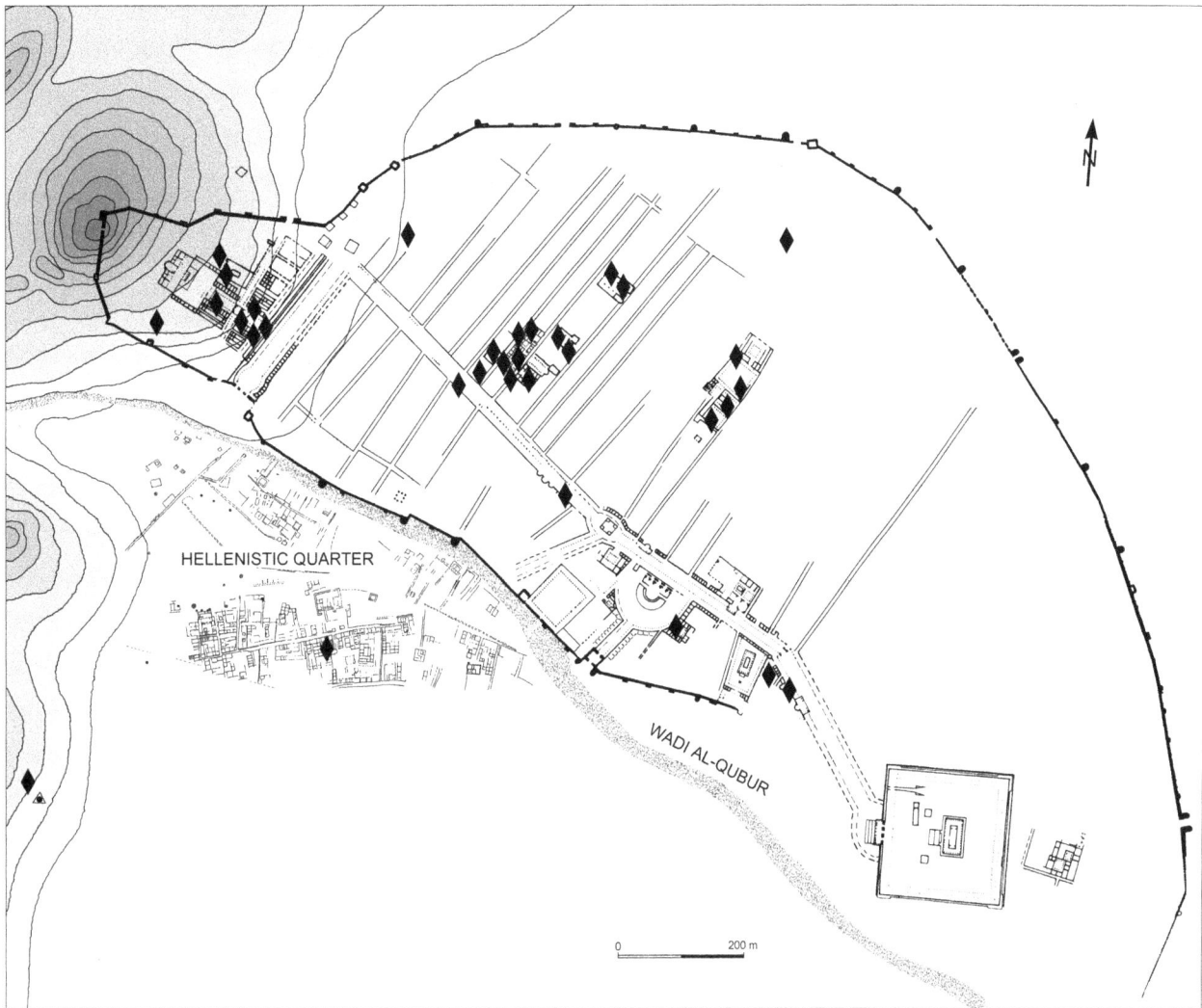

HELLENISTIC QUARTER

WADI AL-QUBUR

0   200 m

Fig 15 Ancient wells documented on the territory of Palmyra

Also the northern aqueduct seems to be in use in this period. During short survey, carried by the authors in 2010, new water line was found between Basilica IV and tower tomb M. Stone-built channel, named Northern Channel, ran on the surface. It is very plausible, that it was connected with the Northern Aqueduct. The chronology of this conduit is uncertain, but stratigraphy suggest rather late period.

In this time water was also collected in different purposes, other than public. Water tanks of total capacity about 100m3 where discovered by Polish mission in the area between Basilica II and Basilica III (GAWLIKOWSKI, 1999, 257). Basins were made of reused column drums and stones. Floors were of bricks set in waterproof, ashy mortar. Water was channeled through two pipelines – one served probably as an inlet second as outlet. Both were connected with the conduits running under the nearby street (GAWLIKOWSKI, 1999, 257).

Basins clearly served for the needs of the churches. It is not clear however, if the water collected was used for drinking, ablutions or industry.Water tanks disappeared together with the churches. This process had been dated by the excavator to IX c. or later.

To Abbasid Period should be attributed another structure, character of which is not clear. It is located in the southern portico of the Great Colonnade, near the north eastern corner of the Theater. Rectangular structure, measuring about 3.70x4.70m and 1.50m high is made of reused blocks originally the architraves of the Colonnade. Inside there is a floor made of bricks set in waterproof mortar. It is quite clear, that this structure was a kind of another water tank. It is located close to one of secondary castellum but not connected with it. If there was some inlet, it was located higher and its not preserved now. Only in the middle of the western wall there is small hole with the remnants of clay pipe inside, probably part of a tap.

The structure is located in the centre of the city, close to the main mosque (GENEQUAND, 2008, 3-15). In our opinion it is very likely that this is feature called *sabil*. In Islamic tradition term *sabil* designate water-house which provided drinking water for free public use. The appearance of such feature in Islamic cities does not seem to be earlier than XII c. (BOSWORTH (ED.), 1995, 679) Chronology of Palmyrene *sabil* is thus uncertain.

Archeology of Islamic water system at Palmyra is still mostly unknown. There are at least three lines of foggaras, underground channels with shafts, which had never been a subject of archeological research and in many publications it is wrongly considered as part of ancient aqueducts. The idea of foggara or qanat came from Iran and was widely spread among Islamic world (LIGHFOOT, 2000, 215-216). It is impossible to establish any chronology of those features at Palmyra, but it seems that some of them were in use until modern times.

As it was said before during the long history of Palmyra wells were used all the time as the supplementary source of water. Many of them are attested inside the houses as well as in the public space, like porticos of the Great Colonnade or temenos of the temple of Baalshamin or Allat. Only few of them were excavated and can be attributed to the specified period. For example the well found by the Polish mission in the southern portico of the western section of the Great Colonnade had to be dug in the second century and in the Byzantine period was already out of use and filled with sand and gravel. But the date of most of them is impossible to determine. They are different in shape – some square, some round, all lined with the limestone blocks and their openings usually measure about 0,5 m. The plan no. 15 shows the disposition of wells on the territory of Palmyra, but the view is surely incomplete and determined by the fact that only small part of the city was excavated till now, so some wells are still buried under the sands of the desert.

Bibliography:

AL AS'AD, KH., SCHMIDT – COLINET, A., 2000
Zur Urbanistik des Hellenistischen Palmyra. Ein Vorbericht, DaM XII, 61 – 93, pl. 7-16

AL AS'AD, KH., STĘPNIOWSKI, F.M., 1989
The Umayyad Mosque in Palmyra, DaM IV, 205 – 217

BARAŃSKI, M., 1997
The Western Aqueduct in Palmyra, Studia Palmyreńskie X, 7 – 17

BOSWORTH, C.E., (ed.) 1995
The Encyclopedia of Islam, New Edition, vol. III

CALLEBAT, L., (ED.) 1973
Vitruve VIII, Paris 1973

CARLE, G., 1923
De l'alimentation en eau de Palmyre dans les temps actuels et anciens, La Géographie XL, VII/VIII 1923, 153 – 160

CROUCH, D., 1975
The Water System of Palmyra, Studia Palmyreńskie VII, 151 - 186

DENTZER, J.-M.; SAUPIN, R., 1996
L'espace urbain à Palmyre. Remarques sur des photographies aériennes anciennes, AAAS XLII, 297 – 318

DASZEWSKI, W.A., KOŁŁĄTAJ, W., 1977
Wstępny raport z wykopalisk Polskiej Misji Archeologicznej w Palmyrze w 1968, Studia Palmyreńskie IV, 69 – 77

DU MESNIL DU BUISSON, R.,1966
Première Campaigne de fouilles à Palmyre, CRAI, 158 – 187

GAWLIKOWSKI, M., 1973
Le temple Palmyrenien Palmyra VI, Warszawa

GAWLIKOWSKI, M., 1974
Recueil d'Inscriptions Palmyréniennes provenant de fouilles syriennes et polonaises recentes à Palmyre, Paris

GAWLIKOWSKI, M., 1999
Palmyra. Season 1999, PAM Reports XI, 249 – 260

GENEQUAND, D., 2008
An Early Islamic Mosque in Palmyra, Levant 40, vol. 1, 3 – 15

HAUK, G.F.W., NOVAK, R.A., 1988
Water Flow in the Castellum at Nîmes, AJA 92, 393 – 407

HODGE, A.T., 1992
Roman Aqueducts and Water Supply, London 1992

HODGE, A.T., 1996
In Vitruvium Pompeianum. Urban Water Distribution Reappraised, AJA 100, 261 – 276

JUCHNIEWICZ, K.J., *et al.*, 2010
The defense wall in Palmyra. After recent Syrian excavations, Studia Palmyreńskie XI, 55 – 73

JUCHNIEWICZ, K.,2010,
Fortyfikacje palmyreńskie. Studium późnorzymskiej architektury obronnej, unpublished Ph.D. dissertation. Conclusions were announced during conference "Palmyra – Queen of the Desert" held in Warsaw 6-8 December 2010.

KOWALSKI, S.P., 1997
Late Roman Palmyra in Literature and Epigraphy, Studia Palmyreńskie X, 39 – 62

LIGHTFOOT, D.L., 2000
The origin and Diffusion of Qanats in Arabia: New Evidence from Northern and Southern Peninsula, The Geographical Journal 166, No. 3, 215 – 226

MATTHEWS, J.F. 1984
The tax law of Palmyra: evidence for economic history in a city of the Roman East. JRS 74, 157-180.

MEYZA, H., 1985
Remarks on the Western Aqueduct at Palmyra, Studia Palmyreńskie VIII, 27 – 33

MILLAR, F., 1993
The Roman Near East, London 1993

PLATTNER, G. A.; SCHMIDT-COLINET, A, 2010
Untersuchungen im hellenistisch-kaiserzeitlichen Palmyra, S. Ladstätter – V. Scheibelreiter (ed.), Städtisches Wohnen im östlichen Mittelmeer 4. Jh. v. Chr. – 1. Jh. n. Chr. Akten des Intern. Koll. Oktober 2007 an der ÖAW Wien, Archäologische Forschungen, Wien, 417-427

PROCOPIUS *DE AEDIFICIIS*
O Budowlach, trans. P.Ł. Grotowski, Warszawa 2006

SCHMIDT-COLINET, A.; AL-AS'AD, KH.; AL-AS'AD, W. 2008
(mit Beitrügen von Ch. Römer-Strehl und R. Ployer)
Unterschungen im Areal der 'hellenistischen' Stadt von Palmyra. Zweiter Vorbericht, ZOrA, 452 – 278

SEYRIG, H., 1934
Antiquites Syriennes, Premiére série. Exstrait de Syria 1931-1932-1933, Paris 1934

STARCKY, J., 1952
Palmyre, Paris 1952

TEIXIDOR, J.,1984
Un port romain du desert, Palmyre et son commerce d'Auguste à Caracalla. Semitica XXXIV

TOURTECHOT, 1735
Les ruines de Palmyre en 1735, published by J. B.Chabot in Journal Asiatique 1897, 335 - 355

WOOD, R.; DAWKINS, H., 1753
Les ruines de Palmyre, autrement dite Tedmor au Desert, Londres

YOUNG, G.K., 2001
Rome's Eastern Trade. International Commerce and Imperial Policy 31 B.C. – A.D. 305, London – New York 2001

ŻUCHOWSKA, M., 2000
Quelques remarques sur la Grande Colonnade à Palmyre. BEtOr , L, 187 - 193

ŻUCHOWSKA, M., 2003
Palmyra. Test trench in the street of the Great Colonnade. PAM Reports XIV, 291 – 294

Notes:

[1] The geophysical prospection followed by the excavations of the Syro – German mission started in 1999 under direction of A. Schmidt Colinet and Kh. al-As'ad and was published until now in preliminary reports only in:AL-AS'AD, KH.; SCHMIDT – COLINET, A., 2000 *Zur Urbanistik des Hellenistischen Palmyra. Ein Vorbericht*, DaM XII, 61 – 93 pl. 7-16, PLATTNER, G. A.; SCHMIDT-COLINET A., 2010, *Untersuchungen im hellenistisch-kaiserzeitlichen Palmyra*, in: S. Ladstätter – V. Scheibelreiter (ed.), Städtisches Wohnen im östlichen Mittelmeer 4. Jh. v. Chr. – 1. Jh. n. Chr. Akten des Intern. Koll. Oktober 2007 an der ÖAW Wien, Archäologische Forschungen, Wien, 417-427

[2] Since the first evidence of the Roman presence in Palmyra comes from the second decade of the I cent. A.D. we will include Augustean relicts also to the Hellenistic period.

[3] Since 1988 Polish Mission is working in the area called Christian Quarter. Until now three churches of considerable size had been excavated and in the fourth the work is proceeding. Annual reports are published in PAM Journal.

NEW EVIDENCE ON ROMAN WATER SUPPLY IN THE EBRO VALLEY: THE ROMAN DAM OF MUEL (ZARAGOZA, SPAIN)

P. Uribe, A. Magallón, J. Fanlo.
University of Zaragoza. Grupo Urbs. Spain.

Key Words: Archaeology, Spain, Ebro Valley, Water supply, Roman Dam of Muel

A new territorial organization unfolded during the reign of Augustus in the Middle Ebro Valley. *Colonia Caesaraugusta* and its road network[1] were not its only material expressions; infrastructure was needed to guarantee the well-functioning of the emperor's new territorial plan.

The Middle Ebro Valley is comprised by several landscapes of varied climates, sharing common characteristics, such as lack of rainfall and alluvial soils with insufficient water flow to supply urban populations. During the Roman period, this shortage becomes manifest in the distribution of water supply systems and related infrastructure, present in every single urban centre of the region.

Roman dams in the Middle Ebro Valley

Reservoirs and dams were mainly located on the tributaries flowing into the right bank of the river Ebro, from the province of La Rioja to Teruel, taking advantage of irregular currents of minor Ebro tributaries, such as the Alhama, La Huerva and Aguasvivas. Other remains of infrastructure destined for urban water supply are found scattered throughout the territory. One of the best and most complex examples of preserved water supply systems are found in *Andelos* (MEZQUÍRIZ, 1988, 237-266) and *Tarraca* (BELTRÁN MARTÍNEZ, 1977, 91-127.), both located on the Ebro's left bank.

Problems arising in research are manifold. Firstly, most works have suffered great destruction and are therefore difficult to identify as Roman. Furthermore, even though Roman hydraulic works are significant in the Middle Ebro Valley, they have not featured accordingly in archaeological literature or Roman hydraulic studies in general, with the exception of two sites, Andelos and Almonacid de la Cuba. This is due to a prior lack of interest in these remains, referenced mainly by public works engineers.

Current studies have sparked interest in hydraulic remains of the Middle Ebro Valley. To traditional studies, focusing on Iberia in general, authored by Fernández Casado (FERNÁNDEZ CASADO, 1983), Arenillas (ARENILLAS, 2003; CASTILLO BARRANCO, ARENILLAS, 2002), Caballero (CABALLERO, SÁNCHEZ-PALENCIA, 1983) etc., we may now add specific approaches to the Middle Ebro Valley, represented by Arenillas (HEREZA, ARENILLAS, CORTÉS, 2000), Beltrán & Viladés (BELTRÁN, VILADÉS, 1994), Dupré (DUPRE, 1997), Mezquiriz (MEZQUÍRIZ, 1979), Martín-Bueno (MARTÍN-BUENO, ALBERTO, 1974., MARTÍN-BUENO, 1975), and Hernández Vera (HERNÁNDEZ VERA, ARIÑO, NÚÑEZ, MARTÍNEZ, 1995), as well as the cataloguing efforts of Arenillas (CASTILLO BARRANCO, ARENILLAS 2002), Fernández Ordóñez (FERNÁNDEZ ORDOÑEZ 1984), and other works veered towards dissemination (ORTIZ, PAZ, 2006,). Two important discoveries of inscriptions, the *Tabula Contrebiensis* (FATÁS, 1980 = CIL I^2 2951a) and the *Lex riui Hiberiensis* (BELTRÁN LLORIS, 2006, 147-197), stress the significance of Roman hydraulics in this region of Hispania. Specific monographs on relevant sites have also addressed issues on water and territory: Ariño's studies on centuriation (ARIÑO, 1986, ARIÑO, 1990), water supply by Vázquez & González (VÁZQUEZ, GONZÁLEZ, 1998), remains from the dam of Almonacid by Beltrán & Viladés (HEREZA, 1996, BELTRÁN, VILADÉS 1994), Almagro's studies of the Cella-Albarracín aqueduct (ALMAGRO GORBEA,

2002), the discovery and recovery of the archaeology of the Jiloca Basin by Rubio Dobón (RUBIO DOBÓN, 2005), and even relevant studies of medieval remains from the Aguasvivas Basin (SESMA, UTRILLA, LALIENA, 2001). All these studies form the background to the work undertaken by the research group URBS (*Grupo de Investigación de Excelencia URBS*) of the University of Zaragoza, which has been studying the Roman dam of Muel, featured in this paper.

The table below shows how several Roman dams of considerable dimensions occupy the central area of the MEV. Their most significant common characteristic is their chronology; the two most important hydraulic complexes, Muel and Almonacid, are dated to the reign of Augustus. C14 dates recorded from sedimentation returned chronologies belonging to the beginning of the first century AD, coinciding with the MEV's great urban development (MARTÍN-BUENO, 1999, 117-122, GALVE, MAGALLÓN, NAVARRO, 2005, 169-214).

The Roman dam of Muel

The dam and reservoir of Muel is an enormously significant site that has gone practically unnoticed until the end of the 20th century. The first published notices date back to 1957 (PELLICER, 1957, 137-146) and the first archaeological study to 1964 (FATAS, 1964, 174-180). Excepting interest shown by several geologists (SILVA AGUILERA) and other specialists (ARENILLAS *et al.*, 2005., ARENILLAS *et al* 2005b, ABADIA, 1995) in Roman engineering, the great dam of Muel remained virtually unpublished[2], even inside the archaeological discipline (LOSTAL, 1980), until the intervention of our research group in 2009.

Location[3]

In its origins, the Roman dam of Muel was located on the river La Huerva[4], although today its course is deviated some metres to the west, due to the silting of the dam in later centuries.

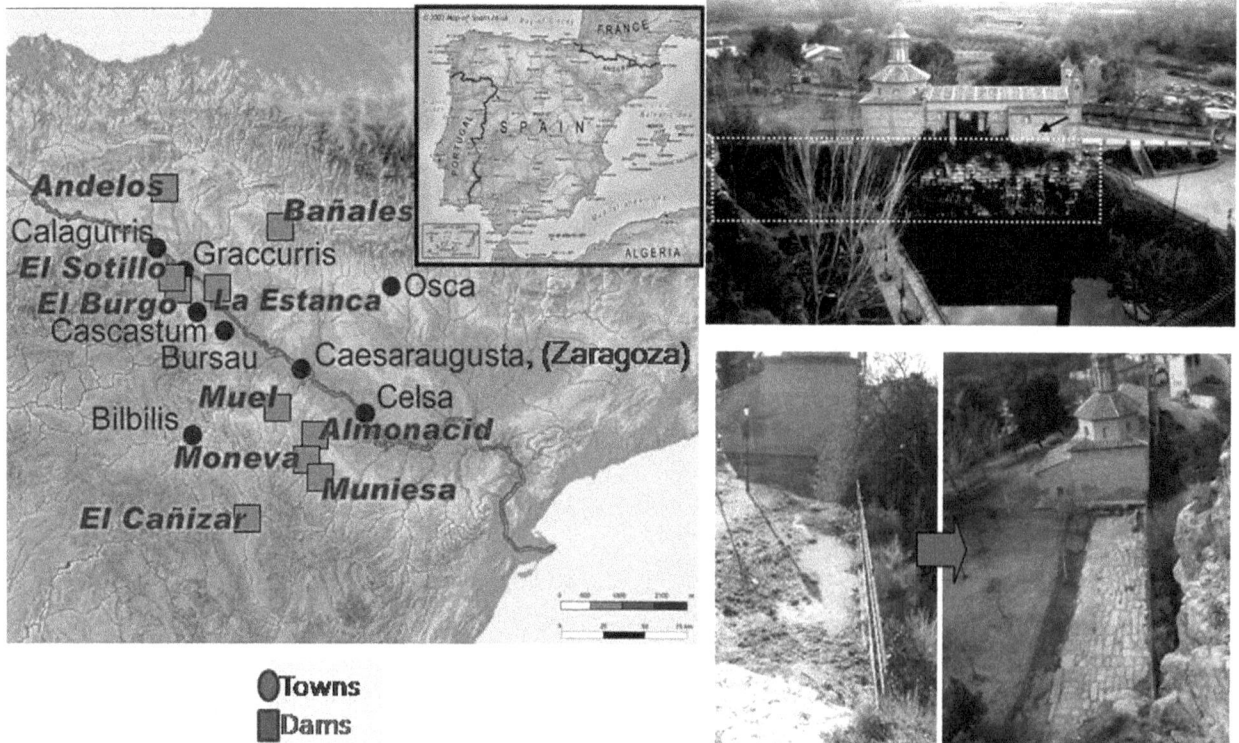

Fig. 1 Location and environs of the Roman dam of Muel

The dam itself was constructed on a narrowing of the river, created by a calcareous headland to the east and to the west by the hill housing the castle of the marquises of Camarasa, not reaching 100 metres in width. Both limestone rock formations easily supported the walls of the dam, creating a solid frame that sustained the pressure exerted by the water. The place chosen by the Roman engineers was one of the few locations that offered the appropriate conditions in the vicinity of *Colonia Cesaraugusta.*

In 1770, the shrine of *Nuestra Señora de la Fuente* was built over the dam, giving the place a special atmosphere. It also informs us of the depth of the dam, which is now the solid foundation to the shrine, decorated in its interior pendentives with paintings by Francisco de Goya.

Size of the dam

As for the possible dimensions of the dam, we relied on an hypsometric analysis of the area located further up the river from the Roman dam, concluding that it could have stretched for some 80 ha. As shown in the aerial photograph, a fertile lowland lies further up the river from the dam, hugged by the eastern bank of La Huerva, and covered by small allotments that cultivate fruit and vegetables, disposed along a perfectly flat and elongated plain, which goes up the river some 2 km x scarcely 400 m.

Technical characteristics of the construction works

The dam survives practically in its entirety, thanks to an early burial. The silting of the duct protected the inner wall from erosion, avoiding the extraction of masonry stone for reuse in later periods, which is exactly what happened with the outer face of the wall –further down the river- or even at the crown of the dam. Test pits from the 2009 campaign[5] uncovered white limestone *opus quadratum*. We had already recorded this technique on a wall further down the river, although in a worse state of preservation. *Opus quadratum* uses carefully carved parallelepiped ashlar, normally placed in horizontal, isodomous or pseudo-isodomous, rows without mortar. The origin of this Roman technique goes back to the Etruscan period (LUGLI, 1957, 245-252 J.P. Adám, 'La construcción 114'), becoming predominant as of the 3rd century BC, although mainly throughout the following one. Therefore, *opus quadratum* architecture took form as Roman rule extended in Hispania, a time of great Hellenistic influences, particularly of Greek construction techniques (ASENSIO, 2006, 117-159).

The excavation reached the water table at 9.35 m., allowing us to record up to fifteen rows of *opus quadratum*, disposed in a stretcher and header bond, in what Lugli calls *maniera romana* (LUGLI, 1957, 181-183, Tav XXXVII). The blocks were made of more or less prominent, polished, dressed stone and a perimetral listel, also called external *anathyrosis*[6]. The dressed stone was smoothed down in two ways: broached[7], with a

chisel or a pointer; or 'pointed', a technique in which the chisel simply flecked off projections in the stone.

Fig. 2 Orthophoto of the dam wall's elevation.

The first Roman evidence for construction bonds using *quadratum* in *Hispania Citerior* (ASENSIO 2006, 133) is quite early. It is recorded on the towers of *Tarraco's* wall, belonging to the end of the 3rd and the beginning of the 2nd centuries BC. In the second half of the 2nd century BC, curtains were built for the second construction phase of *Tarraco's* wall, as well as the wall at *Segeda* II. Likewise, the wall at *Iesso* dates to the end of the 2nd and beginnings of the 1st century BC. Already in the first half of the 1st century BC, the wall at *Iluro* was constructed combining outer faces in *quadratum* with a *caementicium* core. To this brief list of *opus quadratum* construction in *Hispania Citerior*, we must now add the dam of Muel, built towards the change of era during the reign of Augustus (see below).

The masonry unit varied according to its location. The blocks of the first rows were not as tall as those of the last rows. This precaution was taken to lighten the load of walls of considerable dimensions.

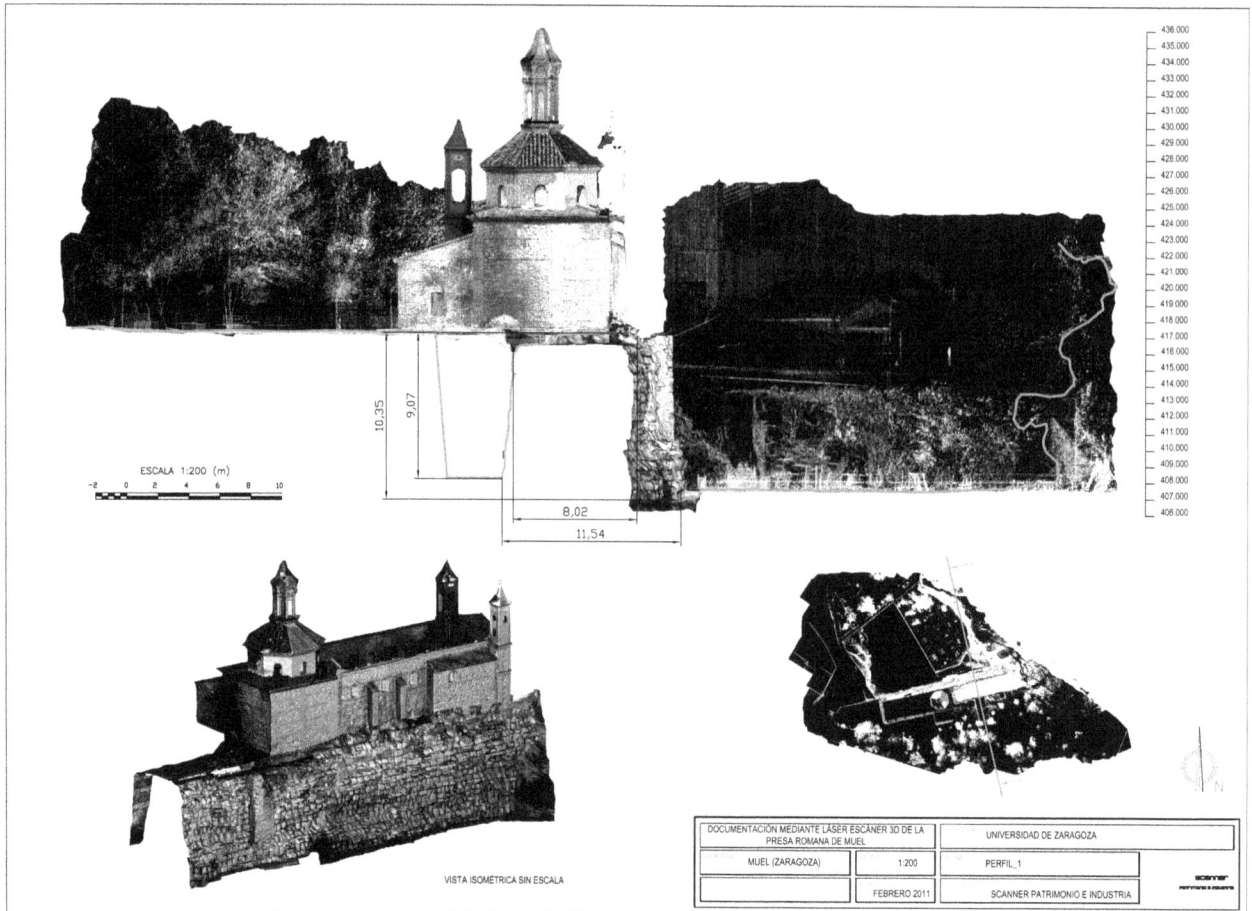

Fig. 3 Cross section of the dam

From the first to the fifth rows, ashlar heights vary between 0.52 and 0.56 m. The taller blocks are located as of the sixth row, oscillating between 0.55 and 0.60 m. As for their width, headers measure from 0.46-0.86 m., while stretchers reach up to 1.5-2.1 m. The width of the perimetral listel varies between 24 and 25 cm. Therefore, we may assert that masonry blocks were modulated according to the Roman foot, 0.296 m., varying in height, between 1.5 and 2 feet (bipedal), as well as in width, between 1.5 and 3 feet for headers and between 5 and 7 feet for stretchers. We already had knowledge on how this modulation worked in similar constructions of *Hispania Citerior,* i.e., those of the second half of the 2nd century BC, such as the walls of *Iesso*[8] and *Segeda* II[9].

The perimetral listel mentioned above is a worked contact edge or fillet of variable width, lining the perimeter of outer faces, typically of dressed stone. Variations in size are significant, and according to Asensio (2006: 148) do not correlate with any chronological criteria. In the Ebro Valley, measurements are known for the wall of *Tarraco* (built c. 200 BC), with 10 and 20 cm. wide listels. Much smaller dimensions are recorded for other cases dating to the mid/end 2nd century BC, such as the 5-7 cm. of the *Segeda II* wall or the 5 cm. of the *sacellum* base from the *Círculo Católico de Huesca* (ASENSIO 2006, 149).

Worth noting are the results returned by particle analyses, which showed that the stratigraphic unit UE 139 was the only non-natural sediment to have silted the dam. UE 139

was composed of sandy gravel from sharp calcareous rock fragments, which according to grain analyses was transported by a great turbulence. This talus, we think, was intentionally created in order to water-proof the works with construction waste, thrown from the top of the dam.

Visible construction joints did not use mortar. Nevertheless, at the crown of the dam, quality *opus caementicium* was used to join the masonry blocks that formed the fill. At this preliminary phase of research, we are still unable to assert whether the dam walls were filled by a compacted mass of *opus caementicium*, or if on the contrary, the interior part of the wall was made in its entirety of ashlar. On the other hand, we do know by analyses that mortar was of good quality –some samples are very similar to Fuller's ideal curve- consisting of a 50-60% main lime composition with a remaining 30% aggregate.

There is no record of the holes left by cramps or traces of scaffolding. The only remaining construction marks are found at the crown of the dam, where small holes, carved into the ashlar, allowed for the placing of the block in its exact position through the force exerted by a lever.

As for the dimensions and structure of the dam itself, the 2009 test pits were able to record that it was a gravity dam without flights of steps –at least at the recorded height-. The slope of the dam reaches a 0.95 m.

inclination in the 8 m. that have been studied. Regarding the maximum height, when adding the row of masonry blocks removed for later reuse, we would be contemplating at least 9.44 m., although we estimate that it could have reached up to 13 m. The width of the dam oscillates between 7/8 metres at the crown and reaches 11.19 m. in the lowest known point. Nevertheless, considering the distance between the two rocky headlands, we think the dam may have reached a length of 100 m.

Chronology

Previous studies (ARENILLAS *et al.*, 2005, ARENILLAS *et al.*,2005b) have always given early dates for the dam, an intuition that was corroborated by the first excavation campaign. The *Legio IIII* marks and C14 dates coincide in dating the Muel dam to the reign of Augustus.

Possible Legio IIII marks

The abbreviation *L(egio)*, followed by an interpunction and the numeral *IIII* was found in test pit 2, six metres down in row 12 of the dam (counting the first pillaged ashlar[10]). The 28 cm. wide inscription was carved on the outer surface of one of the masonry blocks. The height of the letters varies between 12/13 cm., which were executed in one continuous, confident stroke, evidenced by the marks left by the metal tool. The inscribed stone (57 cm. tall x 71 cm. wide) is differentiated from the rest by a small projection, on which the inscription was engraved. It is furthermore remarkable, because it is read backwards, so in order to read it correctly, the block would have to rotate 180°. Its reading could be interpreted[11] as *L(egio) IIII*, for not only is the numeral IIII present, but also the abbreviation for *Legio*, also found on the milestones of Valdecarro in Ejea and Castilliscar (LOSTAL 1992, MPT , no. 18 , 19 and 20).

A new text was inscribed underneath, *P V I,* without interpunction, probably engraved by a different hand. The groove is not as deep and one may even count the number of chisel or punch marks made for each letter. Palaeographically, the characters may be dated to Augustus. As for their interpretation, they may be the initials of a *tria nomina* or letters related with mason marks, although they are different from others found throughout the dam, which are currently being studied and catalogued.

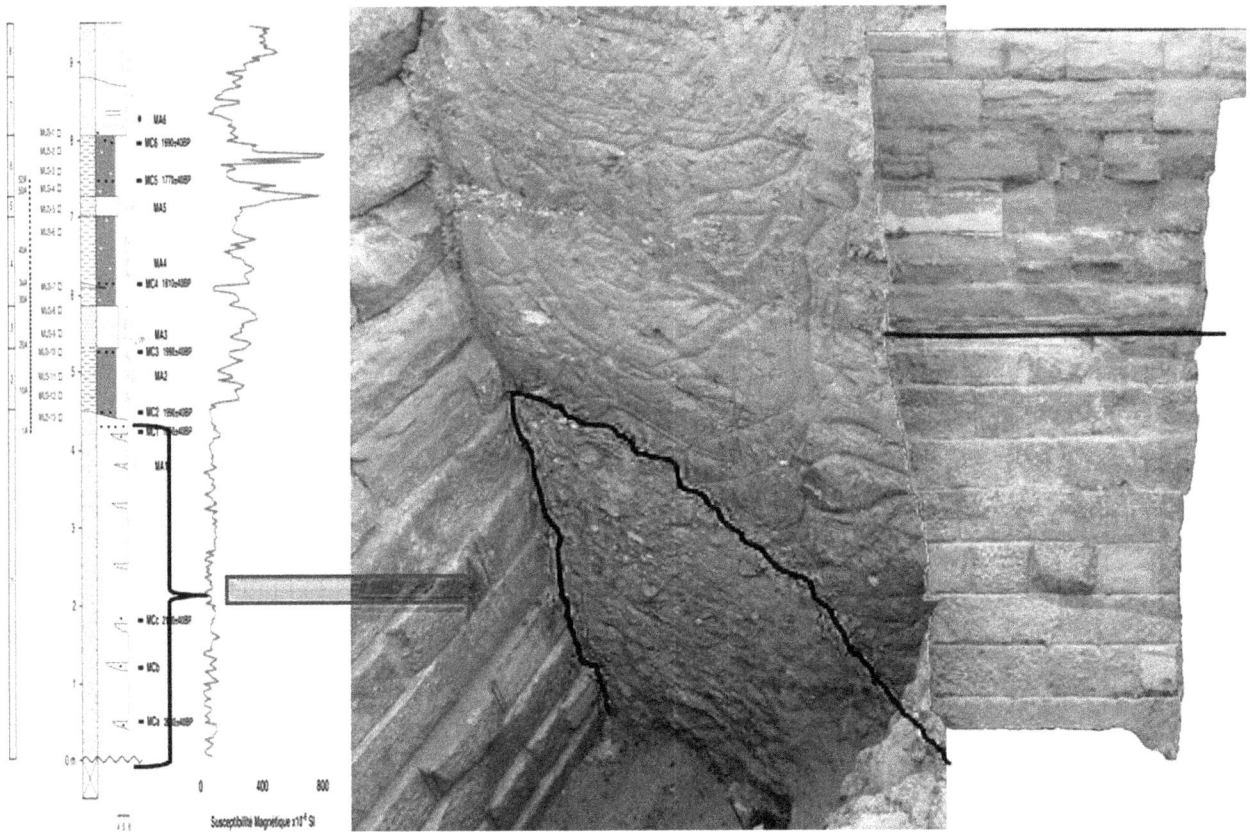

Fig. 4 Water-proofing at the base of the dam

The Roman legions did not only intervene in warfare, but were also very active in processes of territorial organization and the construction of varied public works. Their most relevant contributions in the Middle Ebro Valley were the foundation of the colony and the construction of the Roman road between *Caesaraugusta* and *Pompaelo*. Now we can also say they participated in the construction of the Roman dam of Muel.

The legionaries of the *IIII Macedonia, VI Victrix,* and *X G(emina),* took part in the construction of the Roman road from *Caesaraugusta* to *Pompaelo* (BELTRÁN LLORIS, 1970, 89-117; CASTILLO, 1981, 134-140; AGUAROD,LOSTAL, 1982, 167-218; MAGALLÓN 1987, J. LOSTAL, 2009, 191-238). Between 9/8 and 5/4 BC, members from the colony's three founding legions also constructed the road, according to three of its milestones. One was found in Valdecarro, in the vicinity of Ejea de los Caballeros and two in Castiliscar, inscribed with the names *L(egio) X G(emina), Leg(io) IIII Mac(edonica)* and *L(egio) VI Vi(ctrix)* (MPT , no. 18 , 19 and 20).

Furthermore, the marks of *(Legio) VI* and *(Legio) X*[12] were also found in the so-called Roman port of *Caesaraugusta* and on the reverses of some of the coins minted in the colony (BURNETT, AMANDRY, RIPOLLÉS, 1992 = RPC, Nos. 319 and 326.).

Legionary construction of public works is not only recorded in Muel, but also in one of the administrative landmarks of *Tarraco's* territory, the Roman bridge of Martorell (GURT, RODÁ, 2005, 147-165), which also bares legionary marks.

Fig. 5 Legio IIII mark

For the bridge, marks belonging to three legions have been recorded. The abbreviation *L(egio)* is found seventeen times on the western buttress, sometimes followed by an interpunction and the appropriate numeral: twelve *L(egio) IIII,* three *L(egio) VI* and two *L(egio) X* (GURT, RODÁ 2005, 150).

Carbon 14 dates.

To the chronology given by the inscriptions, we will now add the first C14 results taken from the dam's sediments. These are conclusive. The dam was constructed with Augustus and approximately by the 3rd century it was already buried.

SAMPLE	MATERIAL	DATE	CALIBRATED DATING
Sample MC 1	carbon	1920 +/- 40 1950 +/- 40	40 BC 130 AD
Sample MC 3	carbon	2010 +/- 40 1990 +/- 40	60 BC 80 AD
Sample MC 6	carbon	1650 +/-40 1690 +/- 40	250 AD 420 AD

The dating of samples 1 and 3 give us the moment of construction, which coincides perfectly with the reign of Augustus, while 6 is interesting, because it dates the silting process. In the 3rd century, there are already symptoms of collapse and silting. Currently, a group of geologists, specialists in stratigraphy and palynologists, all part of the Muel work-team, are carrying out sediment analyses, which will no doubt return more data on the silting process of the dam, as well as on the vegetation and climate in the Roman period.

Bibliography:

ABADIA, J,C. 1995
Algunos comentarios sobre el abastecimiento de agua a Caesaraugusta. Cuadernos de Aragón 23.

AGUAROD M. C.; LOSTAL, J. 1982
La vía romana de las Cinco Villas, Caesaraugusta 55-56, 167-218.

ALMAGRO GORBEA, A., 2002
El Acueducto de Albarrarín a Cella (Teruel), Artifex. Ingeniería romana en España, Madrid, 213-237.

ARENILLAS, M. 2003
Presas romana en España Ingeniería y territorio. No. 62, 72-78.

ARENILLAS, M.; CASTILLO, J.C.; HEREZA, J. I.; PINTOR, J. C.; DÍAZ, C.; CORTÉS, R, 2005
La presas romana de Muel en el río Huerva (Zaragoza) IV Congreso nacional de Historia de la Construcción, Cádiz.

ARENILLAS, M.; HEREZA, J.I.;. PINTOR, J. C; DÍAZ, C.; CORTÉS, R.; CASTILLO, J.C.; POCINO, S., 2005b
Caracterización estructural de la presas romana de Muel. Primeros resultados, II Congreso Nacional de Historia de las Presas. Burgos.

ARIÑO, E. 1986
Centuriaciones romanas en el Valle Medio del Ebro: La Rioja, Logroño.

ARIÑO, E. 1990*Catastros romanos en el convento caesaraugustano. La región aragonesa*, Zaragoza.

ASENSIO, J. A. 2006
El gran aparejo en piedra en la arquitectura de época romana republicana de la provincia Hispania citerior: el opus siliceum y el opus quadratum' Salduie, 6, 117-159.

BELTRÁN LLORIS, M. 1970
Notas arqueológicas sobre Gallur y la comarca de las Cinco Villas de Aragón, Caesaraugusta 33-34, 1969-1970, 89-117.

BELTRÁN LLORIS, F., 2006
An Irrigation Decree from Roman Spain: The 'Lex Rivi Hiberiensis', Journal of Roman Studies 96, 147-197.

BELTRÁN LLORIS, F., 2008
Marcas legionarias de la VI Victrix y la X Gemina en el foro de Cesaraugusta, Veleia 24-25, 2007-2008, 1069-1079.

BELTRÁN MARTÍNEZ, 1977
A Las obras hidráulicas de Los Bañales (Uncastillo. Zaragoza) Simposio Segovia y la arqueología romana. Barcelona, 91-127.

BELTRÁN M.; VILADÉS, J., 1994
'Aquae Romanae' Arqueología de la presa de Almonacid de la Cuba, Boletín del Museo de Zaragoza 13, 126-293.

BURNETT, A.; AMANDRY M.; RIPOLLÉS, P.P., 1992
Roman provincial Coinage I. From the death of Caesar to the death of Vitellius (44 BC-AD 69) London-Paris(= RPC). Nos. 319 and 326.

CABALLERO L., SÁNCHEZ-PALENCIA, F.J. 1983
Presas romanas y datos sobre poblamiento romano y medieval en la provincia de Toledo Noticiario Arqueológico Hispánico 14, 379-425.

CASTILLO, C. **1981**
Un nuevo docummento de la Legio IV Macedónica en Hispania, I Reunión Gallega de Estudios Clásicos, Santiago de Compostela (1979), 134-140.

CASTILLO BARRANCO, J.C.; ARENILLAS, M.
Las presas romana en España. Propuesta de inventario I Congreso nacional de Historia de las presas. T.1, Mérida 2002, 199-226.

DUPRE, N., 1997
Eau, ville et campagne dans l'Hispanie romaine. À propos des aqueducs du bassin de l'Ebre Caesarodunum 31, 715-743

FATAS, G., 1964
Nota sobre el dique romano de Muel, Caesaraugusta, 174-180.

FATÁS, G. 1980
Contrebia Belaisca (Botorrita, Zaragoza). II. Tabula Contrebiensis. Monografías Arqueológicas, Zaragoza = CIL I^2 2951a

FERNÁNDEZ CASADO, C. 1983
Ingeniería hidráulica romana. Madrid

FERNÁNDEZ ORDOÑEZ J.A. (dir), 1984
Catálogo de noventa Presas y azudes Españoles anteriores a 1900, Madrid.

GALVE, M.P.; MAGALLÓN, Mª A, NAVARRO, M., 2005
Las ciudades del valle Medio del Ebro en época julio claudia, IV Colloque Aquitania. L'Aquitaine et l'Hispanie septentrionale à l'époque julio-claudienne. Organisation et exploitation des espaces provinciaux, Burdeos, 169-214.

GARCES, I; MOLIST, N; SOLIAS, J. M., 1989,
Les excavacions d'urgencia a Iesso (Guissona, La Segarra) *Excavcions d'urgenacia a les comarques de Lleida*, Barcelona,

GURT J.M.; RODÁ, I. 2005
El Pont del Diable. El monumento romano dentro de la política territorial augustea, AespA, 78, 147-165.

HEREZA, I. (coord.), 1996
La presa de Almonacid de la Cuba. Del mundo romano a la Ilustración en la cuenca del Aguasvivas. Madrid.

HEREZA, I.; ARENILLAS, M.; CORTÉS, R., 2000
Las presas de la cuenca del Aguasvivas. Dos mil años de regulación fluvial, I Congreso Nacinal de Historia de las presas, 55-67.

HERNÁNDEZ VERA, J. A.; ARIÑO, E.; NÚÑEZ J.; MARTÍNEZ, J. M. 1995
Graccurris. Conjuntos monumentales en la periferia urbana: puentes, presas y ninfeos, Logroño.

LOSTAL, J. 1980
Arqueología del Aragón romano. Zaragoza.

LOSTAL, J. 1992
Los miliarios de la provincia tarraconense (conventos tarraconense, cesaraugustano, cluniense y cartaginense), Zaragoza = MPT.

LOSTAL, J. 2009
Catálogo de miliarios de la vía de las Cinco Villas y del Pirineo aragonés, [in:] I. MORENO (ed.), Item a Caesarea Augusta Beneharno, Zaragoza, 191-238.

LUGL, G., 1957
La técnica edilizia romana, con particulare riguardo a Roma e Lazio. Roma, 245-52. J. P. Adám, 'La construcción 114'.

MAGALLÓN, Mª. A., 1987
La red viaria romana en Aragón. Zaragoza.

MARTÍN-BUENO, M., ALBERTO, F., 1974
Análisis de argamasas romanas I. Cisternas de Bilbilis Actas de las I jornadas de Metodología aplicada de las ciencias históricas. Vol.1, 207-14.

MARTÍN-BUENO, M., 1975
Dique romano en Cinco Villas Miscelánea Arqueológica, 251-7.

MARTÍN-BUENO, M., 1999
La ciudad julio claudia, ¿Una estrella fugaz?' II. Congreso de Arqueología Peninsular, Zamora 1996, Vol. 4, 117-22.

MENCACCI, P., 2001
Lucca. Le mura romane. Lucca.

MEZQUÍRIZ, Mª. A., 1979
El acueducto de Alcanadre-Lodosa, Trabajos de Arqueología Navarra 1, 139-47.

MEZQUÍRIZ, Mª A., 1988
De hidráulica romana: el abastecimiento de agua a la ciudad romana de Andelos, Trabajos de Arqueología Navarra 7, 237-66.

ORTIZ, E., PAZ, J., 2006
La vida corriente de las aguas en el Aragón romano. Trabajos públicos y placeres privados Aquaria: Agua, territorio y paisaje en Aragón, Zaragoza, 95-123.

PELLICER, M., 1957
Informe-diario de una prospección por el río Huerva Caesaraugusta, 9-10, 137-46.

RUBIO DOBÓN, J. C., 2005
Contexto hidrogeológico e histórico de los humedales del Cañizar, Zaragoza.

SCHNITTER, N., 2000
Historia de las presas. Las pirámides útiles, Madrid.

SESMA, J. A., UTRILLA J. F., LALIENA, C., 2001
Agua y paisaje social en el Aragón medieval. Los regadíos del río Aguasvivas en la Edad Media, Zaragoza.

SILVA AGUILERA, C.
Catchment scale erosion and sedimentation assessment from valley incision and sedimentation in Roman and present day reservoirs (Central Ebro basin, Spain). The sedimentation of the Muel Roman Dam. Forthcoming doctoral thesis.

SMITH, N.A.F., 1970
The Heritage of Spanish Dams, Madrid.

URIBE, P., FANLO, J., MAGALLÓN, Mª A., 2009
Informe de la campaña de excavaciones del año 2009 en la presa romana de Muel, Salduie IX (in press).

VÁZQUEZ I., GONZÁLEZ, I., 1998
El abastecimiento de agua romano a Caesaraugusta, Anas 1, 35-65.

---

[1] Of the milestones found on the VME, seven belong to Augustus, three of which were signed by the founding legionaries of the colony. MAGALLÓN, 1987 and more detailed in LOSTAL, 1992 (= MPT).

[2] In fact, there is no reference to it in any catalogue featuring Ancient dams (SMITH, 1970; SCHNITTER, 2000).

[3] Data obtained from the *Confederación Hidrográfica del Ebro* (Ebro Hydrographic Confederation).

[4] The river Huerva is about 128 km in length, keeping quite predominantly a north-south direction. Its hydrographic basin carries water from 1000 km^2, which is not such a large area in comparison with other tributaries of the Ebro's right bank (its main counterparts further up and down the river, the Jalon and Martín, have drainage basins amounting to 9338 and 2110 km^2 respectively). This is due to the lack of significant tributaries that would substantially increase its catchment area.

[5] Financed by the *Diputación General de Aragón* (Regional Government of Aragon), URIBE, FANLO, MAGALLÓN, 2009

[6] According to ASENSIO 2006, 148, following Lugli, G., "la anathyrosis tuvo en la arquitectura clásica un origen meramente funcional. Esta lista o filete perimetral se disponía en principio en las superficies de contacto entre los sillares para posibilitar un mejor ajuste entre los mismos, de tal forma que el resto de la cara se dejaba rehundido. No obstante, ya en algunas obras griegas esta franja se traslada a las caras externas de los bloques con el objetivo de enmarcar o limitar el almohadillado, de ahí que a este elemento se le denomine también anathyrosis externa."

[7] The technique "consiste en un alisamiento más o menos cuidado de las superficies externas alternando franjas creadas por marcas oblicuas del cincel o del puntero en sentido derecha o izquierda, de forma que dibujan o imitan la disposición de las espigas de cereal" (ASENSIO, 2006, 150).

[8] On the Iesso wall (GARCES, MOLIST, SOLIAS, 1989, 108-24), the width of the sandstone measured approximately 0.30, 0.44-0.48, 0.58-0.59 and 1.3, which is equivalent to 1, 1.5, 2 and 4.5 Roman feet respectively (ASENSIO, 2006, n.64).

[9] At Segeda II, the ashlar vary in dimensions, but always in multiples of Roman feet, so 1.35 x 0.6 x 0.45 m are 4.5 x 2 x 1.5 Roman feet and 1.35 x 0.6 x 0.3 are 4.5 x 2 x 1 ft (ASENSIO, 2006, n.64). Important Italian constructions of the Republican period using *opus quadratum* show a tendency to use units of 2 x 2 ft for headers and 4/5 ft for the stretchers; such is the case for the walls at Lucca (MENCACCI, 2001) or of Rome's Tabularium (LUGLI, 1957, 197).

[10] The preserved remains seem to indicate that there is a missing row on the upper part of the dam.

[11] Let us recall that in the inscriptions attributed to the founding legions of the colony, discovered in the so-called river port of Caesaraugusta, there is no abbreviation for *Legio,* only the numeral.

[12] Studied by BELTRÁN LLORIS, 2008, 1069-79. These marks correspond to the numerals X and VI, and do not carry any indication of

a *Legio*, despite having been interpreted as belonging to the marks of the colony's founding legions. An early study also interpreted the mark of the IIII *Macedonica*, although this reading was later dismissed.

WASTE AND RAIN WATER EVACUATION IN MEDIEVAL MUSLIM TOWNS: THE CASE OF AL-ANDALUS

Ieva Reklaityte
University of Saragossa
Investigation group Urbs

Key Words: Archaeological and Written Sources; Islamic cities; Al-Andalus; Water; Hydraulic works; Sewers.

Within this paper we do not pretend to elaborate a detailed work on the hydraulics in the medieval Islamic world but to sketch some considerations on the topic by means of the written sources and the archaeological data. It is worth noticing that the inhabitants of al-Andalus as well as of other medieval Islamic territories had to struggle against the soil aridness and lack of potable water during centuries, therefore the access to water was one of the most important concerns of the governors and common citizens.

It is surprising just how thoroughly al-Bakri described water sources as cisterns, rivers and wells, along with the quality and taste of their water in the cities and the transit zones of North Africa (AL-BAKRI, 1965). The work of one of the most renowned intellectuals of the 11th century Spain proves that the presence and easy access to water was one of the most important facilities for medieval traveller, especially in the desert areas. Meanwhile the importance of water in the daily life of the medieval Islamic towns is revealed by means of public and domestic hydraulic installations. Medieval Spain was not an exception in the context of Islamic countries as its inhabitants constructed and repaired hydraulic installations leaving an important legacy after the doom of the Islamic dominion. The importance of the Islamic hydraulic enterprise in Spain can be observed just after the fall of Grenade city in 1492 when a German traveller Hieronymus Münzer observed that apart from being experts in laying out water channels, "The Saracens are very skilled in constructing aqueducts" (MÜNZER, 2002, 79; 81).

Channelled water and fountains

The same traveller observed the presence of channelled water within the houses of the well-to-do citizens of Grenade: "Also the nobles and the rich Saracens possess in Grenade magnificent and famous houses with atriums,

gardens, channelled water and other things" (MÜNZER, 2002, 111).

Nevertheless, the traveller was even more impressed after his visit to the Alhambra palaces where water was present everywhere: "There is so much beauty within the palaces, with their water conducts distributed with ability all over the place that there could be nothing more admirable. From a huge mountain channelled water is conduced by means of a conduct and distributed all over the castle" (MÜNZER, 2002, 95).

Within palatine residences in all the Medieval world, water installations sometimes achieved great monumentality and exceptional beauty. Leon African pointed out that in the town of Fez a watercourse crossed the fortress in order to serve up the necessities of the governor (LEÓN AFRICANO, 2004, 249). An Egyptian 'Abd al-Basit, during his travel along the Grenade Kingdom in the years 1465-1466, was really impressed by the beauty of the fortress of Malaga due to the water presence in it: "In this fortress I saw a hydraulic construction where three large porcelain vessels were present. I have not seen anything like that or similar to it and neither have I heard about something comparable to it. These three vessels were placed one next to another within the construction designed for potable water placed in the vestibule of the fortress and every vessel had a dimension of a tigar or a huge jabia of our country [...] and they were beautifully elaborated and delightfully decorated with admirable and rare embossed adornments" (GARCÍA MERCADAL, 1952, 254).

It must be remarked that in the majority of Medieval urban environments, the presence of channelled water within private houses was a luxury achievable only by the wealthiest citizens.

Benjamin from Tudela narrated that in Damascus of the 12th century houses of the nobles and the public fountains had running water (1989, 84): "The River of Abana [Barada] descends throughout the center of the city and the water is conduced by means of aqueducts to all the kinds of noble houses, to the streets and bazaars". Also Benjamin of Tudela observed the presence of subterranean conducts for water supply to the houses of the aristocracy in the Syrian city of Antioch where a stream "sends water through twenty subterranean aqueducts to the houses of the nobles of the city" (Ibid., 70).

The presence of public fountains must have been frequent within the cities of the medieval Islamic world; nevertheless, the main consumers of its water were the citizens of lower social classes. It seems that a noble was not predisposed to join the common people even when was thirsty. Written sources are quite clear regarding the separation of social classes as it can be observed in a manual for an elegant man written by a Baghdadi al-Washsha' (d. 936). The manual without doubt was intended to be used among the nobles and to show them refined manners and elegancy in all their acts. Among other things the elegant men, such as the author, pointed out (AL-WASHSHA', 1990, 236) "do not drink water from a jug, the water that is sold or water from mosques and public fountains because it is hideous for the discreet people". Through his comments and suggestions it can be observed that the author considered improper drinking and eating freely on the street, a bazaar or a mosque, that means in any place where a congregation of citizens of lower social classes was present.

It must be remarked that not only the residences of well-to-do citizens had current water but the quality and accessibility to water depended on the city sector. Within the morphology of the town of Grenade a social stratigraphy and its relation to the water provision can be clearly observed. The description of the town made by A. Navagero at the beginning of the 16th century suggests the concentration of the citizens depending on their social, economic and even religious adscription within particular city areas. The districts of Albaicin and of Alcazaba were densely inhabited and occupied by tiny dwellings because, according to the traveller, the Moorish "are used to live narrow and crowded" (NAVAGERO, 1983, 50; 132). The same author remarked that the citizens of these districts drank water from the stream of Alfacar which water was extraordinary healthy and almost all the Moorish that continued their habits of eating fruits and drinking only water consumed it. The same author pointed out that all the houses in Grenade had current water that ran through channels that were opened or closed in order to regulate the water flow (Ibid., 51).

Fig. 1. Water cistern. The Fortress of Merida. (Photo by Ieva Reklaityte)

Public Works

Building a mosque, a bridge or digging a public well are, according to Ibn 'Abdun, "acts whose award is treasured in the hands of God" (GARCÍA GÓMEZ, LEVY-PROVENÇAL, 1992, 96). For this reason some governors made large-scale hydraulic works, however, the nobles also financed the installation of wells and cisterns in the public places in order to gain consideration of their citizens. According to N. Hentati, the role of the nobles and their pious donations was very important in the provision of water in the urban environment of the medieval Islamic world (HENTATI, 2001, 171).

Some of the sovereigns were remembered because of their financial support in building hydraulic installations for the benefit of the citizens. The Umayyad ruler Muhammad I (852-886) was the first, according to Ibn Athir (d. 1233), to start public works for drinking water supply in the city of Cordoba, in addition to building a huge cistern of free access to all the citizens of Cordoba (IBN AL-ATHIR, 1898, 231).
The Qur'anic prescription to satiate the thirst influenced, no doubt, in the proliferation of works for water supply that were carried out as pious acts. We can remember the fountain that was built by 'Abd al-Rahman III in 930 AD in Ecija, where a slab foundation explains the reasons that encouraged the sovereign to carry out the work: "hoping to get a good reward from God" (LEVI-PROVENÇAL, 1973, 662).

The sovereign's contribution to the city water supply was considered an enterprise of great admiration; therefore according to al-Bakri, the honor of having built the channel that supplied the city of Igli belonged to the father of the last Umayyad Caliph of the East (AL-BAKRI, 1965, 306). In the Moroccan city of Kairouan, the Umayyad ruler Abd al-Malik Hisham and other princes ordered the construction of several cisterns to supply water to the inhabitants of the city (Ibid., 59). Similarly, in Kairouan, the sovereigns Aghlabids ordered the construction of fifteen cisterns in order to provide the citizens with water. Also in Fatimid Egypt emirs were remembered because of the wells in the neighborhood of Old Cairo, which excavation they financed (AL-MAQRIZI, 1979, 208).

The sovereigns not only controlled the construction of cisterns, but also they sponsored the installation of underground pipes for water supply. Thus, although the town of al-Mahdiya enclosed 360 large cisterns, there were channels traced throughout the city for water delivery. That hydraulic construction was ordered by the Fatimid ruler, according to al-Bakri, who delivered water from a village near the city. The piped water was carried through channels that reached the cistern adjacent to the Great Mosque and from there the water was transported to the governor's palace through a waterwheel (AL-BAKRI, 1965, 66-67). It is worth mentioning that this episode corresponds not only to the generosity of the sovereign but also demonstrates its own interests in order to supply with water his own residence.

Water pipes

Among the Medieval cities which had a subterranean water supply and an underground sewer system, Fez stands out as an exceptional town because of its network of underground pipes installed during the Almoravid period; the system consisted of the networks placed at different depths - the top network was intended for drinking water while the lower network carried used water. There were workers responsible for the supervision of drinking water networks while other workers were in charge of the sewer system (HENTATI, 2001, 188). According to al-Bakri, in the city of Fez, each resident had in front of his house a garden, a mill and indoor plumbing (AL-BAKRI, 1965, 226).

In other cities of the Islamic orbit, the pipeline network also presented technological complexity that required the workforce specialized in hydraulics. Thus, according to Benjamin from Tudela, in the Syrian city of Antioch, "At the top of the mountain there is a spring and there is one man in charge of the source that delivers water through twenty underground aqueducts to the great houses of the city" (BENJAMIN FROM TUDELA, 1989, 70).

Nevertheless, hydraulic achievements not only made urban life more comfortable but also brought tragedies according to some historians. We should point out that due to the engineering skills the hydraulic installations for water supply were able to reach very important constructive magnitudes, thus during the siege of the Andalusian city of Barbastro in 456 H. [December 25, 1063 to 12 December de 1064] the Christian troops were able to use a water supply tunnel in order to penetrate into the town and conquer it (IBN 'IDHARI, 1993, 188). Also Ibn Fadl al-'Omari (d. 1349) narrated that the conquest of the city of Tlemcen by the Marinid army took place because the attackers cut a channel that supplied water to the besieged city (AL-'OMARI, 1927, 193-194).

Use of lead and noble metals

Although the authors of the medieval medical treatises suspected how harmful to health was the use of water pipes made of lead, this metal was often used along with the noble metals as they enhanced the beauty of the environment while stressing the obvious wealth of the owner; Such exorbitant wealth and beauty was seen in a private palatine bath in Baghdad, according to the testimony of al-Maqqari: "I saw it's waters, it's lattices, it's pipes made of silver and covered with gold while others were uncoated. Some had taps in the form of birds and when the water came out of them, a pleasant noise was produced" (RUBIERA, 1988, 98).

In the Caliph palatine town of Madinat al-Zahra' (Cordoba), according to an anonymous author, there was a water pool with twelve zoomorphic figures, made of gold and silver adorned with precious stones, that discharged water from their mouths which poured into the pool (UNA DESCRIPCIÓN ANÓNIMA DE AL-

ANDALUS, 1983, 173). In a novel The Bath of Ziryab, a Moorish author described a bath that a rich citizen of Caliphal Cordoba fabricated for his wife; among other requests aimed at the workers, there were these ones: "I want to have a bath with four rooms, with lead and copper subterranean pipes that would bring hot water to the cold room and cold water to the hot room. On top of each pipe must be figures with eyes made of crimson glass, others [must be] bird-shaped made of brass that would spout cold water through their beaks, and other [figures] made of glass to launch hot water from their mouths" (RUBIERA, 1988, 99).

In the palatine residences, as in the Alcazar of Cordoba, water was brought by means of lead pipes. According to the chronicle of ibn Musa al-Razi (d. 955), water was delivered to the palace of Cordoba from the mountains close to it through lead pipes and from there water was distributed to other parts of the town. The process was so spectacular that people used to come to see the "wonder" (IBN MUSA AL-RAZI, 1975, 21). Also its is worth to recall another description of that hydraulic device that al-Maqqari called a marvel: "Later on the emirs built in the Alcazar true wonders, extraordinary monuments and beautiful gardens irrigated with water brought from the mountains of Cordoba over great distances, through huge pipes that reached the northern side of the enclosure. Then the water ran through every courtyard through lead pipes and withdrew through the fountains that had different shapes and were of gold, silver and copper, filling the huge ponds, wonderful pools and beautiful zafareches with Roman marble pylons with beautiful pictures" (RUBIERA, 1988, 122; AL-MAQQARI, 1840, 208).

The reuse of hydraulic constructions

The magnitude of the Roman hydraulic legacy was admired by the medieval historians, as it is reflected in the narratives left by al-Idrisi, al-Bakri and al-Himyari. The remains of the Roman aqueduct of Miracles in Merida, for instance, inspired a fable about the palace of the legendary Queen Marida (AL-BAKRI, 1982, 34-35; AL-IDRISI, 1974, 171; AL-HIMYARI, 1963, 351-352).

However, some of the Roman hydraulics not only served as an inspiration for legends, but were still in use as some medieval geographers observed.

The testimony of al-Idrisi should be mentioned, since according to him, in the town of Almuñécar a Roman aqueduct was still in use and supplied the city with water: "a square building that looks like a column: broad at bottom and narrow at top" (AL-IDRISI, 1974, 38).

As some of the Roman aqueducts were reused by Muslims in the Iberian Peninsula, the same happened in other areas of the medieval Islamic world. Ibn Fadl al-'Omari (d. 1349) mentions a repaired Roman aqueduct that delivered water to the town of Tunisia and its Great Mosque (AL-'OMARI, 1927, 111-112). The water that supplied the palace of the Sultan of Fez was driven from a remote location through an aqueduct of Roman origin (Ibid., 140). The same author mentions a large cistern in the city of Tlemcen that was also built by the Romans (Ibid., 192).

Fig. 2. Roman aqueduct of Almuñécar reused during the Medieval period. (Photo by Ieva Reklaityte)

In the city of Damascus at the end of the reign of the Umayyads, there were 146 water pipes running. One of them, called Qanawat in fact was a Roman drain which was still in use during the Islamic period (IBN 'ASAKIR, 1959, 255).

The inhabitants of Ceuta, according to al-'Omari (d. 1349), supplied public baths with sea water using water wheels (AL-'OMARI, 1927, 196). This description may also involve an unusual use of a conduct of Roman origin.

Not only large scale hydraulic structures but even water wells could be reused: the prolonged use of wells is documented in Merida, where the Roman wells were still in use during the Visigoth dominion and during the Islamic rule (ALBA, 2004, 435).

Water quality

The water source was the main cause of water quality, as described by various medieval authors. Thus 'Abu Marwan' Abd al-Malik b. Suhr (Avenzoar) (d. 1161-1162) begin_of_the_skype_highlighting considered that water from the cisterns located inside the house in order to fetch water for domestic use, was clearly prejudicial under the healthiness point of view because standing water created a noxious environment that corrupted the moods and promoted the occurrence of fever (PEÑA et al., 1999, 103). The best water, according to Avenzoar, was from "the sources whose birth is oriented toward the sunrise, which, when in contact with heat, heat up quickly or, if they come in contact with cold, cool at the time" ('ABU MARWAN, 1992, 93, 130). Avenzoar believed that the use of current water depended on the season, thus, "is excellent in summer, in winter it is not advisable …" (Ibid., 137).

In the treatise written between the year 1362 and 1371 the famous politician, historian and physician Ibn al-Jatib from Grenade, thus defined various kinds of water: "The best kind of water is from a natural source, especially from those of a warm ground, because its dust has no odd quality. But not every spring sprouts from warm dust, as it must have a continuous course and, especially, if it receives the action of sun and wind […]. Are also good [waters] from upland and those of sweet taste, low weight, rapid cooling or heating, cool in winter and warm in summer, tasteless and odorless, easy to digest and of quick cooking".

When choosing highest quality water, Ibn al-Jatib advised this way: "First, rain water, and in particular that of summer rain. Secondly, water of storms; unless they can quickly become infected due to low texture, resulting therefore prone to the effects of earth and air, but it can be improved by cooking". At the same time, the author suggested not to use water from wells, since their water "is bad because of its condensation, it contains soil and is prone to decompose by the delay. The worst and the most damaging is [water] that flows through the canals and

pipes of lead" (AL-JATIB, 1984, 140). The author stressed that all standing water is harmful to drink, as they are harmful muddy waters that pass through mines or water from a dirty melted snow (Ibid., 141).

River water

The consumption of river water was very common in different Islamic cities as well as in al-Andalus. Thus, the geographer Ibn Hawqal in the tenth century stated that the inhabitants of the Andalusian towns of Calatrava and Malagón caught the water from the river (IBN HAWQAL, 1971, 69). Out of al-Andalus it was also frequent the use of river water: according to Ibn 'Idhari, river water in the city of Tarudan (Morocco kingdom) was of exceptional quality and extremely suitable for consumption: "[...] this river, which water is heavy in sugar, which is clean and digestible, which this beautiful country flowers with [...] "(IBN 'IDHARI, 1954, 328).

However, most of the rivers that passed through the city were exposed to pollution as they also served as a city main sewer. Not in vain in the city of Qasr al-Ifriqi, only the inhabitants of the upper part of the city drank water from the river because it was cleaner at that part of the stream: "The river that runs beneath the city is exploited by the inhabitants of the upper district and they drink its water"(IBN HAWQAL, 1971, 39). In the hisba treatise written in Seville by Ibn 'Abdun contemporary of the Almoravids the muhtasib among other duties had to prohibit the pollution of the river, "[68] should be prevented to through filth and littering onto the riverbank. Do it out of the city gate, on the fields, gardens or places designated for this purpose and not near the river" (GARCIA GOMEZ, LEVI-PROVENÇAL, 1992, 109).

Without doubt, this paragraph reveals that the practice of throwing waste directly into the river was widespread in Almoravid Seville, but not only there. The river also served as the main sewer in the cleanup of the city of Fez, according to Ibn Abi Zar', where the abundant waters dragged the filth accumulated within the city: "The river divides [the town] into two halves and it splits into it by means of the streams, canals and ditches that surround the houses, gardens, estates, squares, markets and baths, move their mills and leave the city, dragging its filth, rubbish and dirt" (IBN ABI ZAR', 1964 , 66).

However, the consumption of river water depended on where it was picked up and on the water quality of other water resources within the urban environment. Al-Zuhri (d. between 1154 and 1161), for example, emphasized that the sweetness of water from the Ebro River was due to the force of the river current (BASSET, 1904, 644). Even physicians sometimes recommend using water from the river: thus for the preparation of eye drops Ibn Wafid (d. 1074) recommended using rainwater collected in a clean vessel, although that kind of water was not suitable during hot weather or when the atmosphere was rarefied. In that case, according to the physician from Toledo, spring water or river water could be used or, in general, any water that met conditions of purity and was sweet [III, 74] (Ibn Wafid, 1980, 83). Also it results surprising the frequent use of rain water in the composition of many of the medications recommended by Abulcasis (d. 1013), that would affirm the purity that was considered to be proper of rain water in the medieval Islamic world (ARVIDE, 2000).

Other water sources

Nevertheless, we should point out that despite the indications of physicians on water quality, common people used to drink water according to its availability and their own tastes. Thus al-'Omari (d. 1349) mentioned that the Tunisian population drank water from wells, while they stored rain water in domestic cisterns and used it to wash clothes and for "other necessities" (AL-'OMARI, 1927, 112). Al-Idrisi when describing Malaga, narrated that its residents drank water from wells as water was almost at ground level, and it was abundant and sweet (AL-IDRISI, 1974, 191).

In the case of Elche, according to al-Idrisi, people used wells despite of having a river nearby, because its water was not drinkable, "Elche is a town built on a plain crossed by a derived river channel. This channel passes under the walls; the inhabitants make use of it, because it's apt for the ships and run through the markets and streets. The water of the river, which is quoted here, is salty. People have to collect rainwater for drinking and to conserve it within cisterns" (Ibid., 183). In another Andalusian town, Niebla, according to the same author, the inhabitants used water from the sources located outside the town (Ibid., 167). In Silves (Portugal), the citizens drank water from a stream that flowed through the city (Ibid., 168).

In the opinion of André Bazzana, reproduced in his work on the cisterns in the Islamic strongholds of al-Andalus, the water that was stored there only served to perform ablutions and for cleaning but not for human consumption (BAZZANA, 1999, 373). Rare are the examples of the Islamic strongholds that would not had at least one cistern, although it was very common to have more than one, including the watchtowers or guard towers (Ibid., 376). The importance of cisterns within a fortified site is reflected, for example, in the fortified area of El Castillejo (The Guajares, Grenade) where there was no hydraulic structures documented inside the enclosure apart from a drain (House 8) and a cistern that must have served as the only water source during a siege during the 13th and 14th centuries (BERTRAND et al., 1990, 207-228).

Although the physicians considered that water from the cisterns was not suitable for human consumption in some areas these were the only water sources. According to the Calendar of Cordoba during the month of December the rainwater was stored in cisterns, because during that month and the next one the water would not taint (LE CALENDRIER DE CORDOUE, 1961, 184). Benjamin from Tudela in his travels in the second half of the

twelfth century noted that the inhabitants of Jerusalem drank mostly rain water collected in domestic cisterns (BENJAMIN FROM TUDELA, 1989, 78).

Fig. 3. Caliphal cisterns from the Fortress of Almería. (Photo by Ieva Reklaityte)

It must be remarked that the presence of domestic wells can be observed in various cities of al-Andalus, while the installation of cisterns probably needed skilled workmen and also they were more expensive. Nevertheless, there are various domestic cisterns documented within the cities of al-Andalus.

In the Islamic Murcia, water wells are documented in almost all the houses, normally installed in the courtyards, kitchens and latrines, although there was documented the presence of several wells in the same house as well (Unit 1, Silver Street, 31 , 33, 35) (RAMÍREZ, MARTÍNEZ, 1999, 553-554). The proximity of artesian level (at only four or five meters from the surface) facilitated the installation of such structures. It is hard to tell if well water was used only for domestic cleaning and for washing or it was also used for human consumption.

Also in medieval Ceuta there are documented cisterns situated under the house floor, as it was observed in the Cervantes Street or in the houses 1 and 2 of the Huerta Rufino settlement. In the case of housing 2 (Huerta Rufino), the cistern had two cameras and a second mouth opened in an auxiliary room (HITA, VILLA, 2000a, 29;

2000b, 303). However, it can't be considered as a common domestic hydraulic structure as it must de stressed that the house was characterized by the presence of decorated rooms and floors, a kitchen of very considerable dimensions, apart from the possible identification of the room with the second mouth of the cistern as a room for ablutions, that would indicate a certain status of their owners.

In the suburb of El Fortí of Denia, beneath the floor of one of the houses there was documented a cistern of great capacity with a square extraction mouth (SENTÍ et al., 1994, 280). In spite of the existence of the cistern, it must be remarked that in the city of Daniya and its suburbs, the location of wells in the courtyard of the houses can be considered as one of the most characteristic aspects of its urbanism. Even the presence of very irregular ground and solid rocks in some areas didn't stop the drilling of domestic wells (GISBERT, 1994, 251-252).

In addition to domestic cisterns we can mention that some public baths were supplied with water from cisterns. The existence of a cistern, for example, is known in the baths discovered in Madre de Dios Street (Murcia) and supplied the installation with water (RAMÍREZ, MARTÍNEZ, 1999, 556). Also water wheels associated with public and private baths were recorded in Murcia, in the Plaza de las Balsas, San Nicolas Street, Pine Street and the Convent of Santa Clara (ROBLES FERNANDEZ et alii, 2002, 544-545). However, not all the baths of Murcia used water from cisterns as the Bath of the Queen (Baños de la Reina), for example, was supplied with water from the nearby canal Aljufía without any need of major hydraulic facilities (RAMÍREZ, MARTÍNEZ, 1999, 555).

The case of the Andalusian Cordoba results paradigmatic in analyzing water supply according to the social position of its citizens. As in other cities, indoor plumbing was only available for the most distinguished residents; water was delivered to the governor's palace, the Almunia of al-Na'ura, the palatine city of Madinat al-Zahra' or the Great Mosque. Lower classes consumed water from wells excavated in the courtyards of the houses while the cisterns must have been quite rare, documented, for example, in some sectors of the western suburbs of Cordoba (plot 1 Polígono de Poniente) (RUIZ NIETO, 1999, 110) (Houses of Naranjal) (CAMACHO et al., 2004, 216). Also water from public fountains must have been widely used; Ibn al-Faqih al-Hamadhani (d. 903), when describing Cordoba, mentioned the abundance of springs and fountains, which provided fresh water and that remained cool in summer (IBN AL-FAQIH, 1949, 53).

Ablution houses

Furthermore we should point out the importance of the mosques in the urban water provision as public water fountains, cisterns or pits normally were situated near them. In the beginning of the 15th century in Ceuta, according to al-Ansari, there were twenty-five public fountains, although the biggest and the most famous fountain was located in front of the entrance to the Great Mosque and was adorned with conduits made of bronze (VALVÉ, 1962, 414; 426). Apart from the fountain, there were two cisterns in every courtyard of the Great Mosque of Ceuta, as al-Bakri informs us (AL-BAKRI, 1965, 202). In the city of Ronda, where due to its topographical and geological characteristics, there was no possible access to the river water, and the construction of a cistern or a well was extremely difficult and therefore most probably quite expensive, a public cistern was located between the Great Mosque and the citadel (AGUAYO et al., 2001, 422-428). Therefore a public well was placed in a strategic urban place, in fact a center of the city, where the most important urban activities (religious, administrative, most probably economic as the bazaar must have been in close proximity) took place.

Apart from the public wells or fountains located near the mosques that provided water to the citizens, the mosques needed water for such essential edifices as ablution houses where religious prescription of self-cleaning before the pray was performed. According to the hisbah treatise, written by Ibn 'Abdun in Almoravid Seville, at least three workers were needed for the maintenance of the Great Mosque: two for its cleaning and lighting and one for water carrying. Also various horses were required for water transportation (GARCÍA GÓMEZ, LEVY-PROVENÇAL, 1992, 84). It must be pointed out that the ablution houses were built not only contiguous to the mosques but also to the religious schools - madrasahs. Al-Ansari relates that in Ceuta there were twelve ablution houses, one of them near a religious school: "It encompassed various rooms and a pool in the middle. The most beautiful and well made is that of New Madrasah, which includes eight rooms and a vast basin for cleansing. In every room there is a marble tub and a bronze drain that fills the tub with water. The pavement is made of carved stone and there is a basin tiled with colorful ceramics [...]. The water is channelled using hydraulic wheels" (VALVÉ, 1962, 426-427).

Nevertheless, a unique description of an Andalusian ablution house was made by Hieronymus Münzer, a German traveller , who visited Grenade in 1494 and still observed the characteristics of the Muslim town that Grenade had been before the Christian conquest occurred:

"Outside that mosque there is an edifice, in the centre of it there is a large marble basin twenty feet length, where they wash up before entering the mosque. Around it there are small constructions with water conduits for their latrines and sewers, which consist of an opening in the floor, one cubit length and a hand wide. There is a constant water flow beneath it. There is also a small basin for urinating. Everything is constructed so accurately and nicely that provoke admiration. There is also an excellent well for drinking water" (MÜNZER, 2002, 91).

Although the ablution house described by Münzer hasn't survived, there have been several ablution houses documented archaeologically in al-Andalus as those of Cordoba and Seville.

Archaeological excavations in 1998 in Cordoba (current Magistral González Francés Street) revealed the characteristics of the ablution house adjacent to the Great Mosque that was amplified by al-Mansur (d. 1002) during the Caliphal period although the foundation of the edifice was made by the emir Hisham I (d. 796). (MONTEJO, 1998, 253-255; 1999, 209-235). The magnitude of the ablution house was very notable as the edifice was sixteen meters wide and more that twenty-eight meters length. The edifice was built of massive stone blocks and a complex sewer network was installed beneath the latrines.

In 1994 the ablution house of the Great Mosque of Seville, built in the second half of the twelve century, was unearthed. Although the structure of the edifice bears a resemblance to the one from Cordoba, it was made of "cheaper" materials such as brick and dry mud (tabiya) and the proportions of the edifice were inferior (MONTEJO, 1998, 255; VERA, 1999, 107-109). The main characteristic of the ablution house of the Great Mosque of Seville is its sewer that bordered the entire edifice.

Finally we should remark that the ablution houses were present not only near the mosques or madrasahs but also adjacent to other type of edifices. In the palatine complex of Alhambra, an ablution basin along with the latrines was installed near the Secondary Mexuar or the Royal Chancellery, where the personnel could perform their religious duties (LÓPEZ LÓPEZ, ORIHUELA, 1990, 123).

Waste water evacuation: a privy

In a Muslim world, as H. Mortada resumed it, "The privacy of the home and woman is a vital principle that has been declared in sharī'ah basic sources. […] Since it is a religious principle, the privacy of the individual and his family should be maintained in both houses and neighborhoods alike. […] Islam recognizes the right of every individual to be free from undue encroachment on the privacy of his or her life. Therefore, the privacy of the house is significantly stated in many places in

Fig. 4. Ablution house adjacent to the Great Mosque of Cordoba. Photo by Ieva Reklaityte.

the Qur'an" (MORTADA, 2003, 78; 83; 95). One of the rights that every Muslim boasts inside his house is a visual privacy, and the design aspects of a house are thought to protect the living space from visual intrusions. It can be observed within various rules made by the scholar Malik 'Ibn Anas (d. 795) as that of a vision angle of a person standing at the door must be narrow enough to prevent visual intrusion of the immediate area (e.g., entrance lobby or hall) behind the door of the opposite house (Ibid., 98).

Therefore these principles would be reflected in the traditional Muslim building practices. It can be seen from the archaeological evidence that a room of a latrine is normally placed opened at the patio and near the entrance. Apart from curtains and doors, the privy was often concealed by a low wall that would impede viewing the person that was using the toilet by anyone that would enter the room. Although the latrines in the houses of the Military district of the Alhambra (Grenade) did not possessed twisted entrances, the latrines of the noble houses and the Nasrid palaces had access corridors with one, two or even three bends, doors being absent. (FERNÁNDEZ-PUERTAS, 1997, 64). In the Almohad Palace found beneath the Courtyard of Monteria del Real Alcazar of Seville, it was possible to document that normally narrow corridors, which lead to the principal rooms were placed in central position of the house, meanwhile the accesses to the latrines and kitchens were placed obliquely (TABALES, 2001, 237).

Therefore inside the house, privacy and intimacy were key considerations whilst the latrine was being used. We can refer to the oriental Islamic world, where in the course of a didactic book designed for the high classes of the Baghdad society in order to teach them how to comport themselves in a sophisticated manner, al-Washshā' (d. 936) states that an elegant man by no means enters a latrine while someone is watching him, nor relieves himself in company (AL-WASHSHĀ', 1990, 273).

It is therefore extraordinary to encounter evidence of several double latrines within Islamic dwellings. An example of a latrine where two platforms were installed – in order to be used by two persons at the same time – is documented in the so-called Building of Services used by the servants of a dignitary in the royal town of Madinat al-Zahra' (Cordoba) (VALLEJO TRIANO, 1990, 131). Another example of a double latrine has also been found in tenth-century Cordoba, this time in one of the dwellings of the inner city (RUIZ NIETO, 1999, 108). Theoretically, only the lower classes of the society would have shared the latrines, relieving themselves in presence of other persons. We can only suppose that these double latrines were used by slaves whose individual need for intimacy was of little concern to their owners.

Normally, a latrine was an extremely reduced space opening into the patio and located near the entrance. The major part of the latrines that are documented archaeologically can be considered as rather uncomfortable because of their smallness. We can

remember a privy excavated in Murcia (Yesqueros Square-Toro Street), which belonged to the House 1 that was only 75 cm wide (ROBLES, NAVARRO SANTA-CRUZ, 1999, 582); or the latrine found in the Courtyard of Oranges of the Cathedral in Seville, built between the end of the 11th and beginning of the 12th centuries, which was 60 cm wide and 1 meter length (JIMÉNEZ SANCHO, 2003, 905-922).

A small rectangular platform made of brick or stone with a narrow opening in the middle was placed inside; a person would have had to squat in order to relieve himself. Most likely slim apertures in the upper part of the wall were used to ventilate the room in conjunction with the air that would circulate from the patio1. Although the latrine was considered to be dirty and unpleasant, it was at the same time the most intimate space of the house, and some of the latrine rooms were decorated with mural paintings and exquisite tiling. Naturally these houses were owned by quite affluent citizens in a position to afford not only the decoration of principal rooms but also the adornment of the latrine. Al-Hamadhani (d. 1007), a Persian writer, described this way the latrine of a wealthy citizen:

"Thither stucco aloft it
And cistern below it
And ceiling polished above it
And smooth marble spread under it;
An ant might not cling to the sheen of its wall
And a fly on the floor there would slip down and fall;
Its door-panels' joints mixed ivory, teak,
In the finest skilled joining one ever did peek!" (cited by TALBI, 1998, 379-461).

The royal palace of the Alhambra (Grenade) serves as a consummate example of the sophisticated architecture under the Nasrid kingdom that is reflected even in their splendid latrines (AGUILAR GUTIÉRREZ, 1989, 232). Rich mural decorations in lavatories are also found in other remarkable dwellings in Seville (see the archaeological survey in the house of Miguel de Mañara) and Cordoba (Orive Palace) both constructed during the Almohad period (OJEDA CALVO, 1999, 138; MURILLO REDONDO et al., 1995, 181).

Waste water evacuation: sewers and cesspools

By means of the archaeological evidence we can observe that normally the citizens of al-Andalus used to connect their latrines to the cesspits rather that to the sewer network. Apart from the more extensive use of cesspools, a sewerage network was present in some medieval Muslim cities; the presence of sewers is documented through archaeological excavations and even by means of the written sources. It must be said that the installation of a sewer and its maintenance required some engineering knowledge. Also the setting up of a sewer most probably was quite costly and moreover required the contribution of the major part of the inhabitants of the district that would have shared the sewer. Such cities as Murcia,

Lérida, Denia, Málaga, Almería, Cordoba or Algeciras had sewer network of greater or lesser importance (GISBERT, 1994, 251-261; NAVARRO, JIMÉNEZ, 1995, 401-412). Normally, rainwater and residual waters were discharged through the sewers into the river. Nevertheless, the most popular solution of waste water evacuation wasn't the use of sewers but cesspits (NAVARRO, JIMÉNEZ, 2005). Normally the cesspits were located in the public street, adjacent to the façade of the house, although it was common to have them installed in the cul-de-sacs. In the cities where cesspits and not subterranean sewers were used, rain water was never discharged into the cesspools but it was conduced to the public streets, as it was observed in medieval Saragossa, for instance (GUTIÉRREZ GONZÁLEZ, 2006). The recommendation not to connect between cesspit and rainwater channels was quite logical, since cesspools would have overflowed during torrential rains.

Fig. 5. A latrine. Vivienda Superior, Madinat al-Zahra' (Cordoba). Photo by Ieva Reklaityte.

The cesspools had to be cleaned from time to time and there were specific workers in charge of their clean-up as the author of the hisbah treatise Ibn 'Abdun informs us (GARCÍA GÓMEZ, LEVY-PROVENÇAL, 1992, 149). The filth from the cesspits was used as a fertilizer and it seems that in the Islamic cities fecal material was put up for sale as highly-prized manure. This fact can be deduced from various fatwas related to the disputes that arose during this sort of transactions (GARCÍA SANJUÁN, 2002, 177-178).

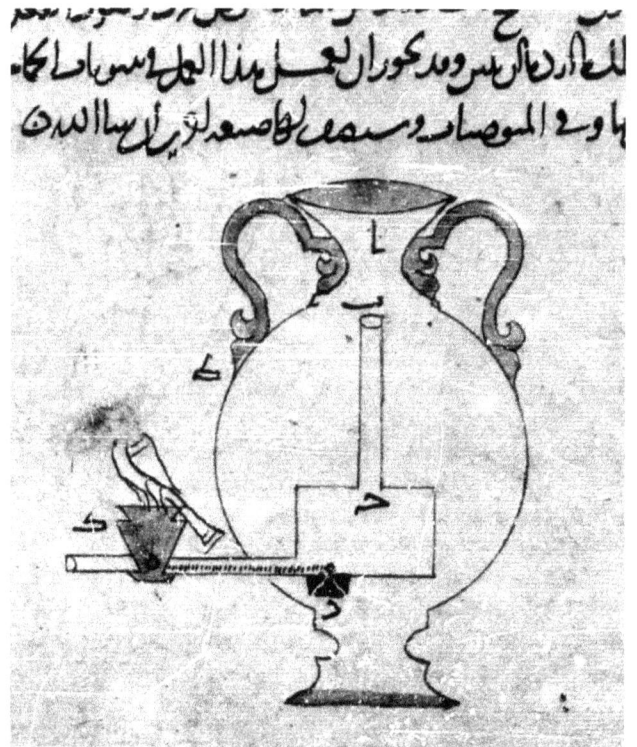

Fig. 6. A water device for ablutions and bathing (BANU MUSA BIN SHAKIR, 1979, 82).

Final notes

Finally we conclude by stressing that the importance of water in the Islamic world is reflected in the abundance of hydraulic constructions and devices for its delivery and discharge. The higher social strata preferred channelled water in their homes not only for hygienic purposes or for human consumption but as part of an architectural decoration. For example, the work of three brothers from Baghdad, called Banu Musa bin Shakir, which was written in the mid-ninth century, contains a number of mechanical devices designed to convert a water source in an admirable set of colored water, make it run hot or cold, rise or fall. These mechanical devices were designed to be placed in the baths or ablution places within private residences; other devices intended for receptions and celebrations mixed wine with water in a surprising way (BANU MUSA BIN SHAKIR, 1979).

Written sources as well as archaeological data reveal the abundance of hydraulic works (cisterns, aqueducts, wells) that were concluded in order to provide the city with water, at the same time some of the Roman aqueducts and canals were reutilized during the Islamic period. The quality and purity of potable water was one of the concerns within the Islamic medical treatises. Some of the most important and costly hydraulic works were ordered directly by the sovereign who expected the gratitude of the citizens and the comfort of himself. Only the most well-to-do citizens afforded to have current water in their houses while the major part of the population used water from private or public wells, cisterns or river sources. Faucets, pipes and decorative figures of the fountains were made of lead and sometimes of precious metals in order to increment the beauty of current water. Although various Andalusian towns presented an extensive sewer network system, the most common solution was a use of cesspits. Also rivers could have served as cloacae in order to evacuate that way part of the town rubbish. The contamination of urban environment and the corruption of potable water were the main problems caused by the deficient waste water and rubbish removal procedures.

Bibliography:

'ABU MARWAN 'ABD AL-MALIK b. SUHR (AVENZOAR), 1992
Kitab al-Agd□iya (Tratado de los alimentos), (translated by E. García Sánchez), Madrid.

AGUAYO DE HOYOS, P., CARRILERO, M., PADIAL, B., 2001
Excavación arqueológica de urgencia en la Plaza Duquesa de Parcent de Ronda (Málaga), Anuario de Arqueología de Andalucía, 1997, Sevilla, 422-428.

AGUILAR GUTIÉRREZ, J. , 1989
Restauración de pinturas murales en la Alhambra. Patio del Harén y Retrete de la Sala de la Barca, Cuadernos de la Alhambra 25, 204-211.

ALBA CALZADO, M., 2004
Apuntes sobre el urbanismo y la vivienda de la ciudad islámica de Mérida, Memoria 7, 2001, Mérida, 417-438.

AL-BAKRI, 1982
Geografía de España (Kitab al-masalik wa-l-mamalik) (translated by E. Vidal Beltrán), Zaragoza

AL-HIMYARI, 1963
Kitab ar-Rawd al-mi'tar (translated by M. P. Maestro González), Valencia.

AL-IDRISI, 1974
Geografía de España (translated by R. Dozy and M. J. Goeje), Valencia.

AL-MAQQARI, 1840
History of the Mohammedan dynasties in Spain (translated by P. De Gayangos), Vol. I, London.

AL-MAQRIZI, 1979
Les marchés du Caire (translated by A. Raymond and G. Wiet), Caire.

AL-'OMARI, 1927, *L'Afrique, moins l'Egypte* (translated by M. Gaudefroy-Demombynes), Paris.

AL-WASHSHA', 1990
El libro del brocado (translated by T. Garulo), Madrid.

ARVIDE CAMBRA, L. M., 2000
Un tratado de oftalmología de Abulcasis, Almería.

BANU MUSA BIN SHAKIR, 1979
The Book of Ingenious Devices (Kitab al-Hiyal) (translated by D. R. Hill), London.

BASSET, R., 1904
Extrait de la description de l'Espagne tiré de l'ouvrage du Géographe anonyme d'Almerie, Homenaje a Francisco Codera, Zaragoza, 619-647.

BAZZANA, A., 1999
"Al-djubb": le stockage de l'eau dans les édifices castraux et les habitats d'al-Andalus", Castrum 5, Achéologie des espaces agraires méditerranéens au Moyen Age, Murcia, 371-399.

BENJAMÍN DE TUDELA, 1982
Libro de viajes de Benjamín de Tudela (1130-1175) (translated by J. R. Magdalena Nom de Déu), Barcelona.

BERMÚDEZ PAREJA, J., 1974-1975
El baño del Palacio de Comares, en la Alhambra de Grenade. Disposición primitiva y alteraciones, Cuadernos de la Alhambra, 10-11, Grenade, 99-117.

BERTRAND, M., CRESSIER, P., MALPICA CUELLO, A., ROSSELLÓ-BORDOY, G., 1990
La vivienda rural medieval de "El Castillejo" (Los Guájares, Grenade), in La casa hispano-musulmana, (J. Bermúdez López, ed.), Grenade, 207-228.

CAMACHO CRUZ, C., HARO TORRES, M., LARA FUILLERAT, J. M., PÉREZ NAVARRO, C., 2004
Intervención arqueológica de urgencia en el arrabal hispanomusulmán "Casas del Naranjal". Yacimiento "D". Ronda Oeste de Córdoba, Anuario Arqueológico de Andalucía, 2001, Sevilla, 210-230.

CRONICA DEL MORO RASIS VERSIÓN DEL AJBAR MULUK AL-ANDALUS DE AḤMAD IBN MUḤAMMAD IBN MUSÀ Al-RAZI, 889-955; ROMANZADA PARA EL REY DON DIONÍS DE PORTUGAL HACIA 1300 POR MAHOMAD, ALARIFE, Y GIL PÉREZ, CLÉRIGO DE DON PERIANES PORÇEL, 1975,
(D. Catalán and M. S. De Andrés, (eds), Madrid.

GARCIA GÓMEZ, E.; LÉVY-PROVENÇAL, E., 1992,
Sevilla a comienzos del siglo XII. El tratado de Ibn 'Abdun, Madrid.

GARCÍA MERCADAL, J., 1952
Viajes de extranjeros por España y Portugal, Madrid.

GARCÍA SANJUÁN, A., 2002
Hasta que Dios herede la tierra. Los bienes habices en Al-Andalus siglos X al XV, Huelva.

GISBERT SANTONJA, J. A., 1994
Daniya-Dénia-. Remembrança d'una ciutat andalusí, IV CAME, Tomo II, 1993, Alicante, 251-261.

GUTIÉRREZ GONZÁLEZ, F. J., 2006,
La excavación arqueológica del Paseo de la Independencia de Zaragoza, Zaragoza.

HENTATI, N., 2001
L'eau dans la ville de l'Occident musulman médiéval d'après les sources juridiques malikites, Revue d'histoire maghrébine, 102-103, Zaghouan (Túnez), 165-220.

HITA RUIZ, J. M., VILLADA PAREDES, F., 2000a *Un aspecto de la sociedad ceutí en el siglo XIV: los espacios domésticos*, Serie minor. Estudios y ensayos, 2, Ceuta.

HITA RUIZ, J. M., VILLADA PAREDES, F., 2000b
Restos de viviendas de un barrio mariní de la Ceuta islámica (tercera campaña de excavación en Huerta Rufino, Q☐urtuba, 5, Córdoba, 301-304.

IBN ABI ZAR', 1964
Rawd al-qirtas (translated by A. Huici Miranda), Valencia.

IBN AL-JATIB, 1984
Libro de Higiene (translated by M.a C. Vázquez de Benito), Salamanca.

IBN 'ASAKIR, 1959
La description de Damas d'Ibn 'Asakir (translated by N. Elisséeff), Damas.

IBN EL-ATHIR 1898
Annales du Magreb et de l'Espagne (translated by E. Fagnan), Alger.

IBN HAWQAL, 1971
Configuración del mundo (fragmentos alusivos al Magreb y España) (translated by M. J. Romani Suay), Valencia.

IBN 'IDHARI AL-MARRAKUSHI, 1954
Al-Bayan al-Mugrib (translated by A. Huici Miranda), Colección de crónicas árabes de la Reconquista, vol. III, Tomo II, Tetuán.

IBN 'IDHARI AL-MARRAKUSHI, 1993
La caída del Califato de Córdoba y los Reyes de Taifas (al-Bayan al-Mugrib) (translated by F. Maíllo Salgado), Salamanca.

IBN WAFID DE TOLEDO, 1980
El libro de la almohada (translated by C. Álvarez de Morales y Ruiz Matas), Toledo.

LE CALENDRIER DE CORDOUE, 1961
(published by R. Dozy, translated by Ch. Pellat), Leiden.

LEÓN AFRICANO, J., 2004
Descripción general del África y de las cosas peregrinas que allí hay (translated by S. Fanjul), Grenade.

LÉVI-PROVENÇAL, E., 1973
Historia de España (R. Menéndez Pidal, ed.), T. V, Madrid.

MORTADA, H., 2003
Traditional Islamic Principles of Built Environment, New York.

MURILLO REDONDO, J. F., CARRILLO, J. R., CARMONA, S., LUNA, D., 1995)
Intervención arqueológica en el Palacio de Orive, Anuario de Arqueología de Andalucía, 1992, Sevilla, 175-187.

MÜNZER, J., 2002
Viaje por España y Portugal (1494-1495), Madrid.

NAVAGERO, A., 1983)
Viaje por España (1524-1526), (translated by A. González García), Madrid.

NAVARRO PALAZÓN, J., JIMÉNEZ CASTILLO, P., 1995)
El agua en la vivienda andalusí: abastecimiento, almacenamiento y evacuación, Verdolay 7, Murcia, pp. 401-412.

NAVARRO PALAZÓN, J., JIMÉNEZ CASTILLO, P., 2005
Siyāsa. Estudio arqueológico del despoblado andalusí (s. XI-XIII), Historia de Cieza, vol. II, Murcia.

OJEDA CALVO, R., 1999
El edificio almohade bajo la casa de Miguel de Mañara," [in:] Sevilla Almohade, (M. Valor Piechotta and A. Tahiri, eds), Sevilla-Rabat,135-141.

PEÑA, C., GIRÓN, F, BARCHIN, M., 1999
La prevención de la enfermedad en el Al-Andalus del siglo XII, [in:] C. Álvarez de Morales and E. Molina, (eds) La medicina en al-Ándalus, Grenade.

RAMÍREZ ÁGUILA, J. A., MARTÍNEZ LÓPEZ, J. A., 1999)
Introducción al urbanismo de la Murcia islámica a través de una intervención de urgencia en los solares número 31, 33 y 35 de la calle Platería (junio-octubre, 1994), Memorias de arqueología, 9, 1994, Murcia, 548-569.

ROBLES FERNÁNDEZ, A.; NAVARRO SANTA - CRUZ, E.; MARTÍNEZ ALCALDE, M., 2002
Sistemas hidráulicos y transformaciones urbanas en el sector oriental de Mursiya. Informe preliminar de la intervención realizada en la Plaza de las Balsas, n. 15, Memorias de Arqueología 10, Murcia, 534-551.

RUBIERA, M. J.,1988
La arquitectura en la literatura árabe, Madrid.

RUIZ NIETO, E., 1999
Intervenciones arqueológicas en el Polígono de Poniente durante los años 1993 y 1994, Anuario de Arqueología de Andalucía, 1995, Sevilla, 104-112.

SENTÍ RIBES, M. A., GISBERT SANTONJA, J. A., BERENGUER LLOPIS, M. J.1994)

L'espai privat al Raval de Daniya (El Fortí. Dénia), IV
CAME, Tomo II, 1993, Alicante, 277-287.

TALBI, M., 1998
Everyday life in the cities of Islam, [in:] A. Bouhdiba,
(ed.), The different aspects of Islamic Culture. The
individual and Society in Islam Paris, 379-461.

VALLEJO TRIANO, A., 1990
La vivienda de servicios y la llamada casa de Ya‹far,
[in:] La casa hispano-musulmana, (J. Bermúdez López,
ed.), Grenade, 199-241.

VALLVÉ BERMEJO, J.,1962
Descripción de Ceuta en el siglo XV, Al-Andalus,
XXVII, Grenade-Madrid, 398-442.

UNA DESCRIPCIÓN ANÓNIMA DE AL-ANDALUS,
1983
(translated by L. Molina), Madrid.

Notes:

[1] The presence of high narrow windows is documented in the latrine of
the bath in the Palace of Comares (The Alhambra, Grenade) that was
used for room ventilation (BERMÚDEZ PAREJA, 1974-1975, 106).

LATE MEDIEVAL CITY OF KALISZ WATER SUPPLY: EVIDENCE OF ARCHAEOLOGICAL DATA AND WRITTEN SOURCES

Tadeusz Baranowski,[a] **Urszula Sowina,**[a] **Leszek Ziąbka,**[b] **Robert Żukowski**[a]

[a]**Institute of Archaeology and Ethnology, Polish Academy of Sciences;** [b] **Regional Museum of Kalisz**

Key words: Poland, Kalisz, Water supply, Archaeology, Written sources, Late medieval period

Kalisz is one of the most important centres of early and medieval Poland. located in the south of Great Poland on the Prosna river (fig. 1.). Its major advantage, both in antiquity and in the Middle Ages, was a convenient location because of communication conditions – the course of trade routes, etc. Kalisz received city rights probably in 1257 and then received privileges upon the German town law. Before that the nucleus of the city was known as the Old Town – an open settlement established at the elevations – islands of the Prosna valley bottom, in connection with a fortified stronghold on "Zawodzie" ("the other side of the water") – the seat of secular and ecclesiastical power) with the church of St. Paul (BARANOWSKI (ed.) 1998; BARANOWSKI 2004, 285-304; BARANOWSKI 2010, 167-178).

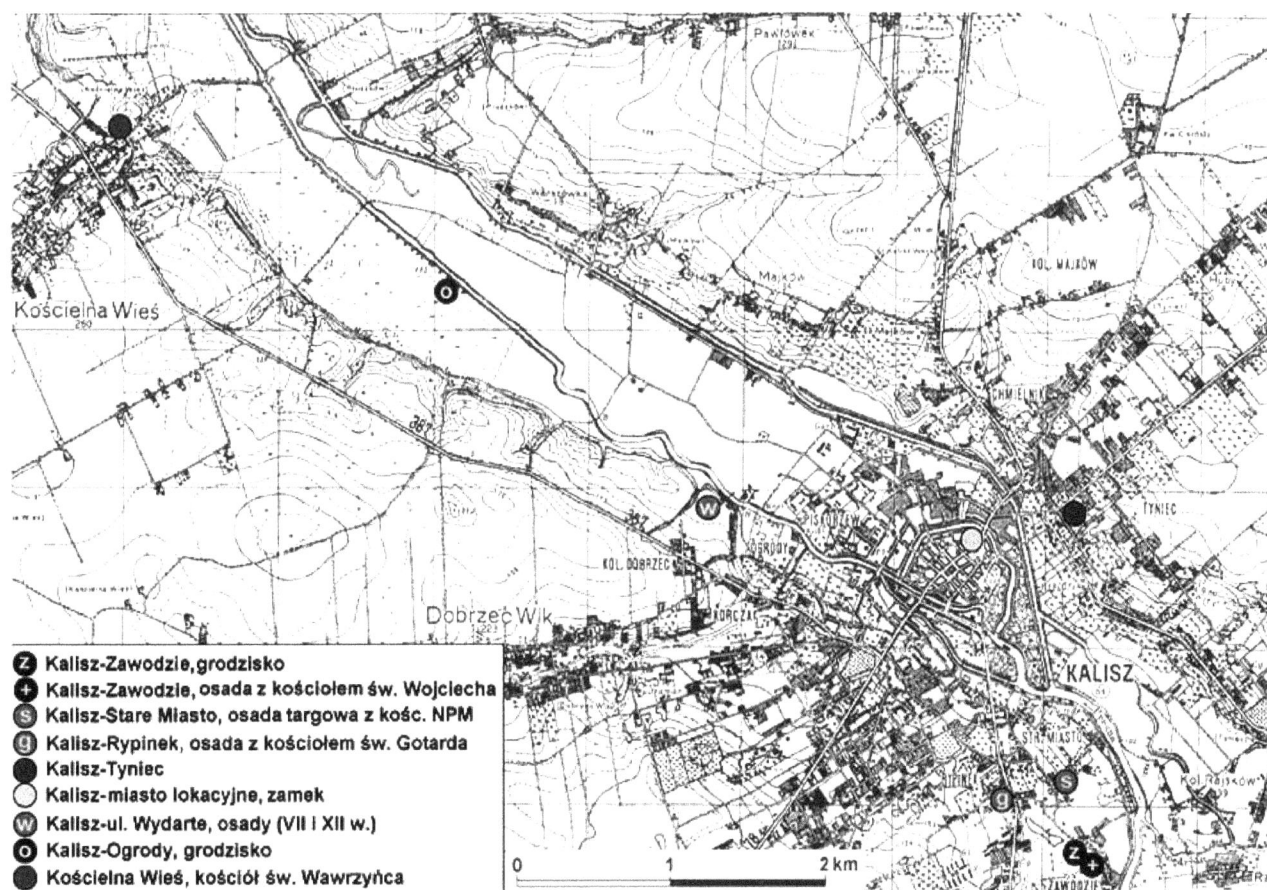

Fig. 1. Topographic map of the area around the town of Kalisz.(z- Kalisz-Zawodzie stronghold). Prepared by T. Baranowski and G. Calderoni

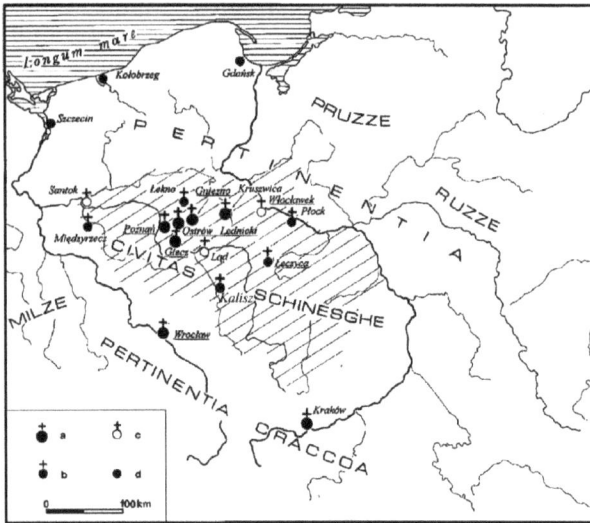

Fig. 2. Map of the first Polish State.

An open settlement nearby, stronghold Zawodzie with the church of the St. Adalbert, also belonged to the early medieval settlement unit, and – on the high bank of the Prosna valley – St. Gotthard church, in whose vicinity was probably the first Jewish settlement, and a little further – the corresponding cemetery. Kalisz hosted one of the first Jewish communities in Poland – except Przemyśl, Wrocław, Kraków (JUREK 1992/1993, 29-53; GRODECKI 1969). The significance of the Jewish settlement is confirmed by numismatic sources of Kalisz (bracteates of the prince Mieszko the Old – from the second half of XII century) (SUCHODOLSKI 1973), and by the Statute for Jews of Kalisz Duke Boleslaw the Pious from 1264, in which Jews were subject to the sovereignty of Prince, and not to the authorities of cities where they lived (WITKOWSKI, 2007; ZAREMSKA, 2005; TOPOLSKI, MODELSKI (eds), 1999).

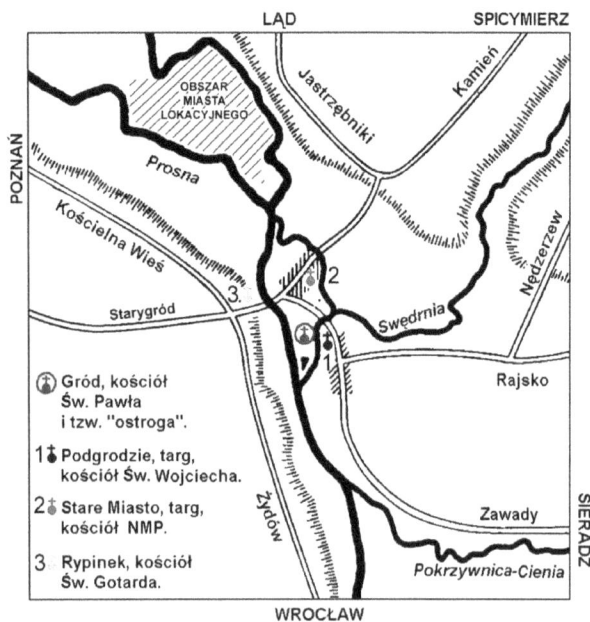

Fig. 3. Map of the Medieval settlement unit of Kalisz.

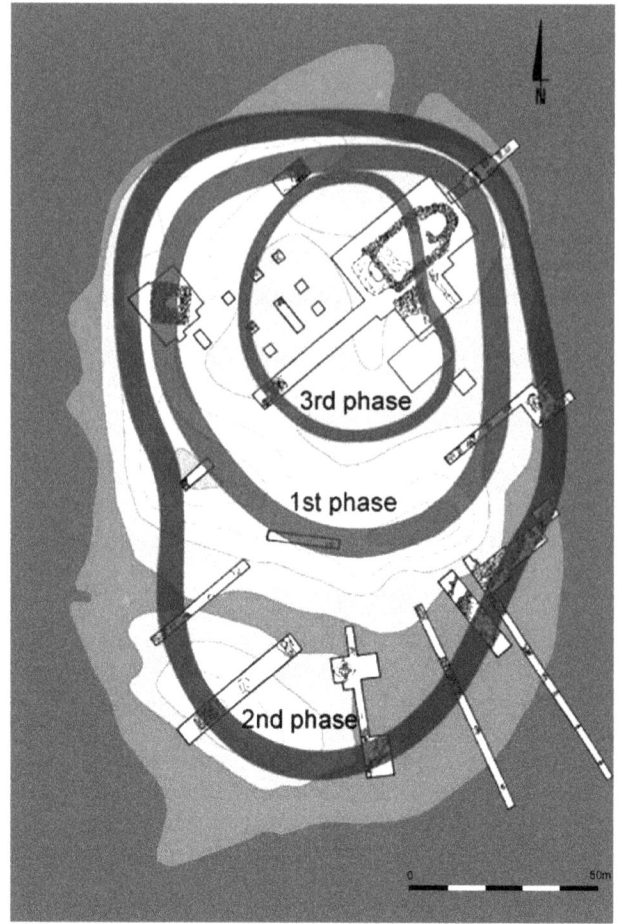

Fig. 4. Three phases of the history of Kalisz-Zawodzie early Medieval stronghold, with position of the archaeological excavations from 1958-1992.

The stronghold in Zawodzie had its heyday in the times of Mieszko III the Old (1122/1125-1202). This ruler used Kalisz in particular as his capital, besides Poznań and Gniezno(M. PRZYBYŁ 2002). In the mid-12th century he founded a stone church – St. Paul's collegiate church. The remains of this temple were uncovered by archaeologists supervised by Krzysztof Dąbrowski in the 1950s and 1960s (DĄBROWSKI, 1976; DĄBROWSKI, 1977).

In the 12th century, most probably because of the constantly rising level of natural water table and danger of floods, the stronghold was made smaller and thus easier to defend. The stronghold became a real centre, retaining the functions of a centre of lay, church, political, and administrative power, and the settlements surrounding it became an early-urban hinterland.

The period between the capture of the stronghold by Henry I the Bearded in 1233 and the location of the town in ca 1257 was relatively frequently mentioned in the written sources. However, there is no certainty among the archaeologists about the geographical situation of the political and administrative centre of Kalisz:, i.e. the duke's residence, Some believe that the ducal *palatium* was the stone tower, the remains of which were found during the excavations of the western part of the highest (latest) ring of defensive walls of the stronghold (TOMALA, 2004). A team of archaeologists and historians of architecture that excavated the stronghold in the 1980s believe that before 1233 the ducal seat was probably in the southern section of the tallest part of defensive walls. However, with respect to later times, these researchers tend to adopt the hypothesis that the centre of lay power was moved to another place, at present impossible precisely to locate.

At present the most important research tasks are excavations in the so-called Old Town. This is the part of Kalisz where the medieval town originated and where a unit which can be called an early town certainly functioned.

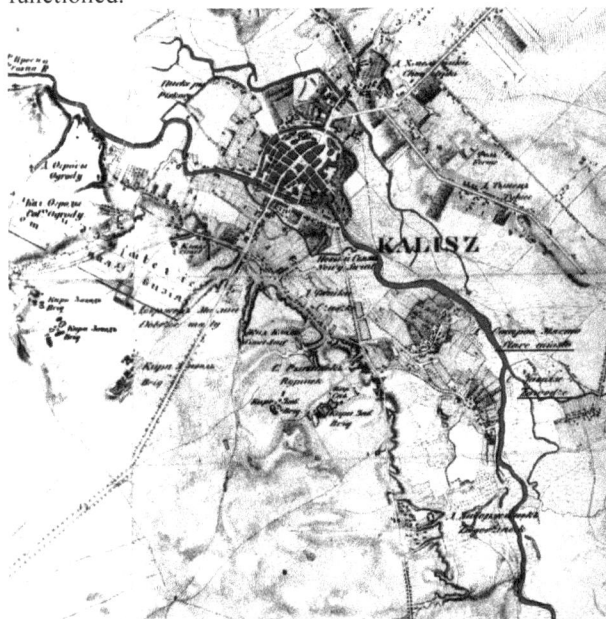

Fig. 5. Map of Kalisz, beginning of the 19. century.

Fig. 6. Marketplace of Kalisz with well. (after *Przyjaciel Ludu* 4, 25.07.1835, 25)

Fig. 7. Marketplace fire and a well with the pump.(Tygodnik Ilustrowany 1866)

The settlement in the Old Town basically fulfilled all the functions of a town (WĘDZKI, 1977). It was an important crafts and trade centre. It held one of the markets and a customs office of the early medieval Kalisz. As it seems, the archaeologists have found part of the market place. At that area, i.a., a considerable number of scale weights was found.

Also traces of a building, perhaps an inn, were uncovered. A small hoard of Duke Bolesław the Curly's (1121/1122-1173) coins was found inside, hidden in a box and also a bone dice. There are other noteworthy finds. In the Old Town a considerable number of coins were found, starting with two fragments of two Arabic dirhams from the 10th century.

The discovery of a small deposit consisting of five Saxon denarii and a fragment of a gold object representing a bull's head is particularly significant in this context. Among other things, on the basis of these coins, it is possible to locate a so-far unknown mint producing coins in the early 12th century. It belonged to duke Zbigniew (1070/1073-1114), son of Ladislaus I Herman and brother of Boleslaus III the Wry-Mouthed (1086-1138) (KĘDZIERSKI,2010). This is a very important discovery, which placed Kalisz among the most important centres of the Polish State. Other mints existed in Kalisz in the times of Mieszko III the Old and, probably, Przemysł II. The importance of the existence of duke Zbigniew's mint in Kalisz is also due to the fact that it functioned in the period before the fratricidal fight between Ladislaus Herman's sons, which was described in the Gallus Anonymous' chronicle together with a mention about Boleslaus III the Wry-Mouthed capture of Kalisz in 1106.

Some bracteates of Mieszko III the Old have Hebrew inscriptions, but it is only the Hebrew alphabet (script): the language is Polish (ZAKRZEWSKI, 1923; SUCHODOLSKI,1973).

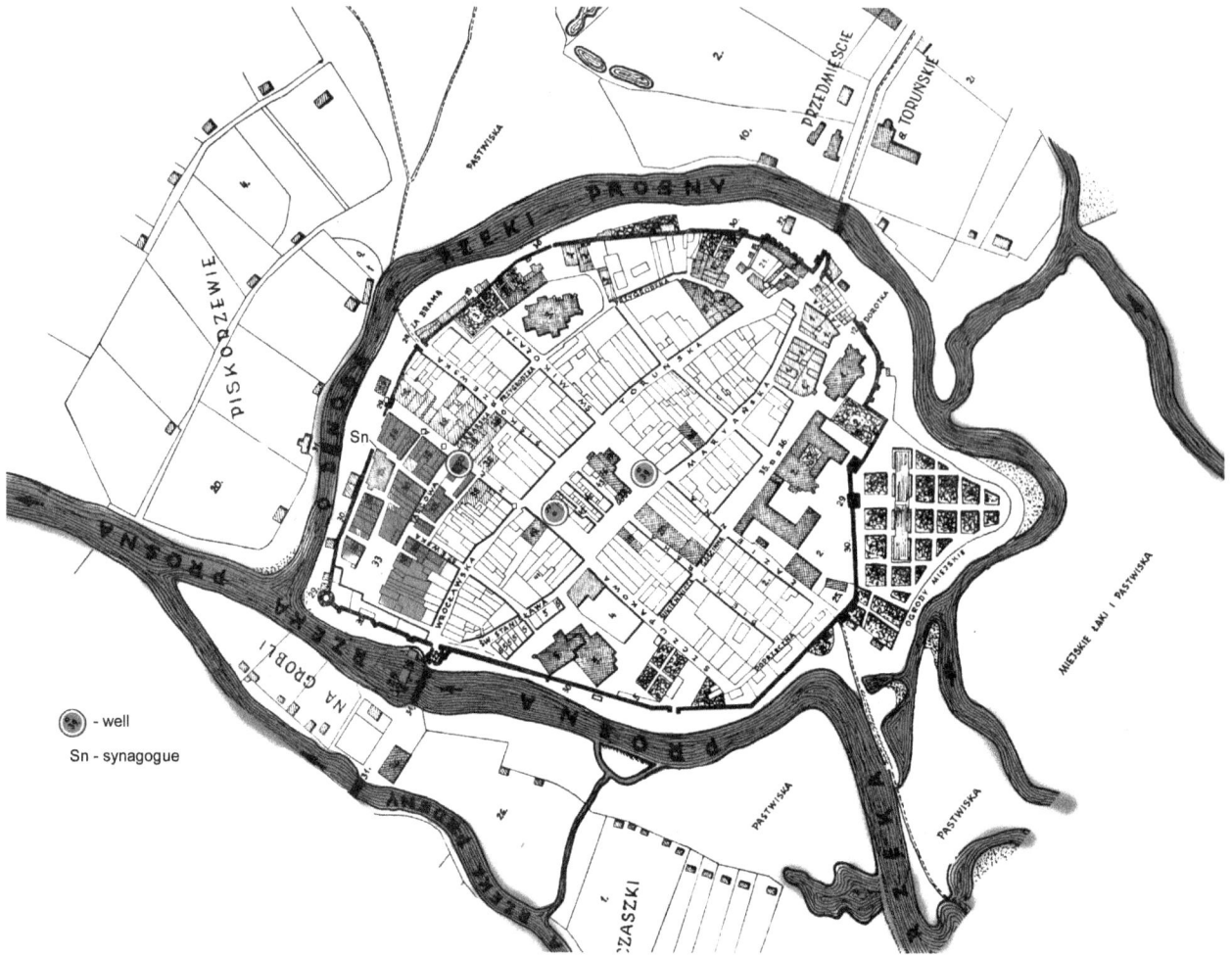

Fig. 8. Map of Kalisz with three of 5 wells – cistens.

As for the process of the rise of the medieval Kalisz, the settlement in the Old Town area has not changed in the medieval town for two reasons – there was no place for the regular layout of the new city, and also because of the flood disasters that plagued this part of the Prosna valley occurring during the thirteenth century, serious consequences of climate changes, responsible for adverse water relations.

It is possible to talk about the early phases of the town only in the context of the stronghold in Zawodzie, but especially about the early urban settlement outside the stronghold, situated in the area of modern Old Town Street, where a location town was not created only because of unfavourable natural environment, namely the dangerous floods and the scarcity of space where the urban area could be planned according to the location principles. Still less suitable was the area of the former stronghold in Zawodzie. Thus the location town was established at the area where the centre of Kalisz is now, several kilometres down the river from the stronghold and settlement in the Old Town. This place provided a suitably large area, was safe from the floods and had excellent conditions for communication and transport. A small difference in of tens of centimetres in ground level between the Old Town and the New City (which was

located slightly higher), was sufficient to protect against flooding the new centre, regardless of that, that was situated on the islands of the river branches. It should be emphasized that the location of the medieval Kalisz, was on the scraps of land in the middle of the water of the Prosna River. At the beginning of the fourteenth century (as we know from the mention from the year 1303) (MŁYNARSKA, 1960, I, 111) through the centre of the city flowed a small stream that separated the two sites inhabited - the two parishes established in the city – that of the church of the St. Nicholas from the other church of the Virgin Mary. The first of these was the seat of the secular authorities – later castle and late medieval Jewish quarter, on the second was situated the other authorities, the ecclesiastical – the collegiate. This is a repetition of the situation with the early medieval stronghold on Zawodzie, where the headquarters of both the authorities were separated by watercourse. The course of the stream coincides roughly with the main course of communication in the city – north-south – from the gate Toruń, to the Wrocław gate. Soon, this area with a stream was buried, and squared. This indicates that the ground level of city ever been subject to the changes, it constantly increased, so has changed the level difference in relation to the former centre of the Old Town.

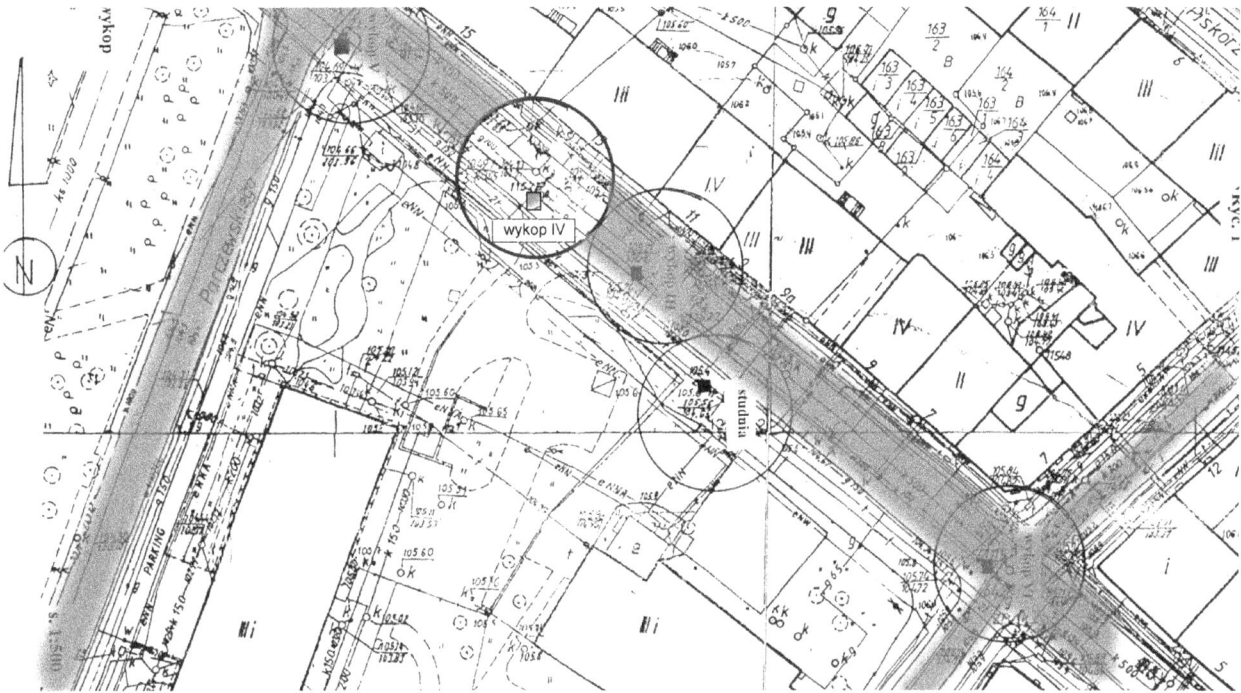

Fig. 9. Area which authors took in consideration: Gold street with the archaeological trenches (IV – with pipes; 2 – with well – cistern).

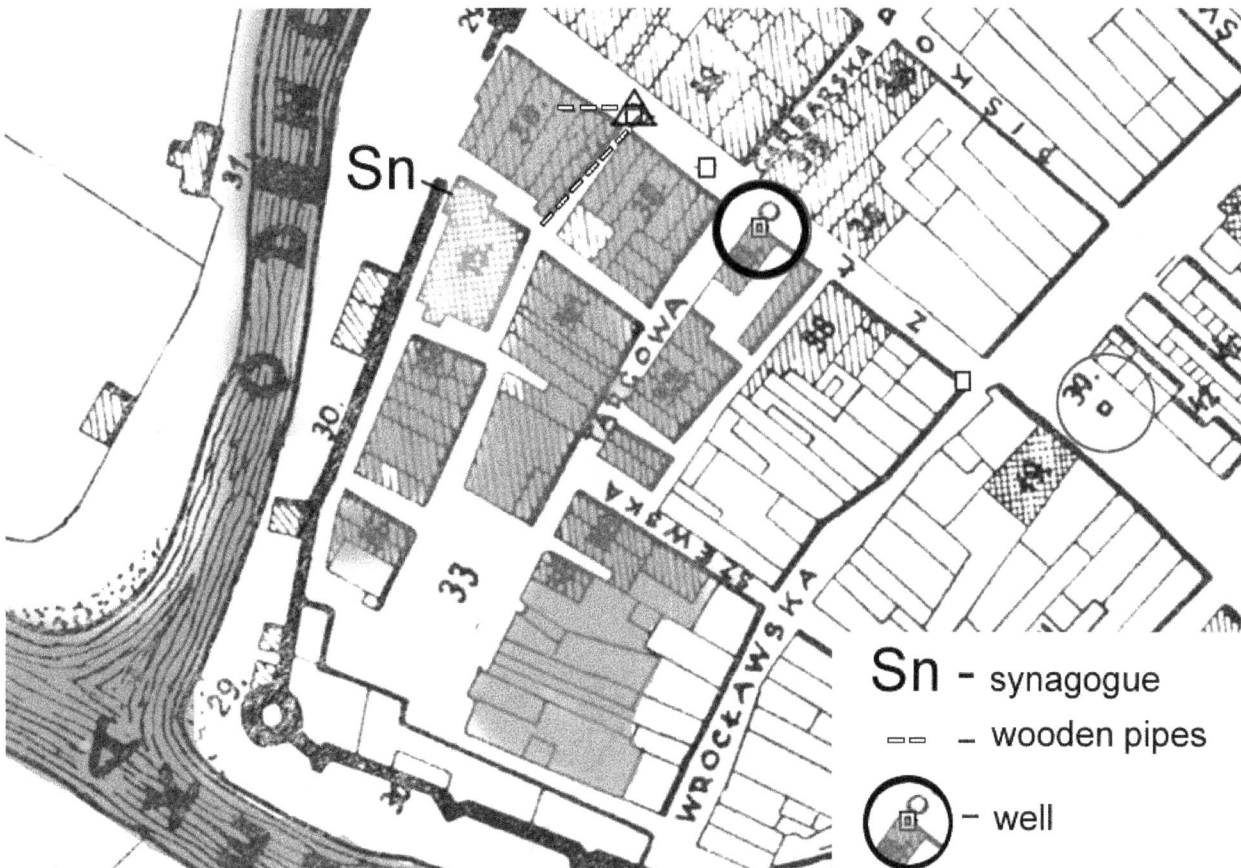

Sn - synagogue

▭▭ - wooden pipes

⊙ - well

Fig. 10 Area of the city of Kalisz with location of two wells, and direction of wooden pipes.

Fig. 11. Fragments of the wooden water pipes *in situ* during the excavations of Gold Street at Kalisz (fot. L. Ziąbka).

suburban area (e.g. in 1394 the mill mentioned in the Old City of Kalisz, which led the way through the role of the pastor of Dobrzec and can it be used only with permission of the Judge of Kalisz) conducted by the land.

The watermills were mainly flour mills, but there were also fulling mills, or at least fulling wheels in flour mills, a forge hammer driven by a mill wheel and a sawmill. It transpires from the sources we have that the king Ladislaus Jagiello issued a decision of its construction on 27 April 1402, motivating it with the needs of the town and the castle alike. After this year one could therefore expect more construction works in those two premises. The sawmill was therefore to serve the „consumption" needs of the locals (following yet another fire, perhaps?) and not the revival of the lumber production for sale on the external market, which would be more profit-generating than the sale of unprocessed trunks (distribution of revenue and expenses from the sawmill was following: 2/3 king's share, 1/3 town share).

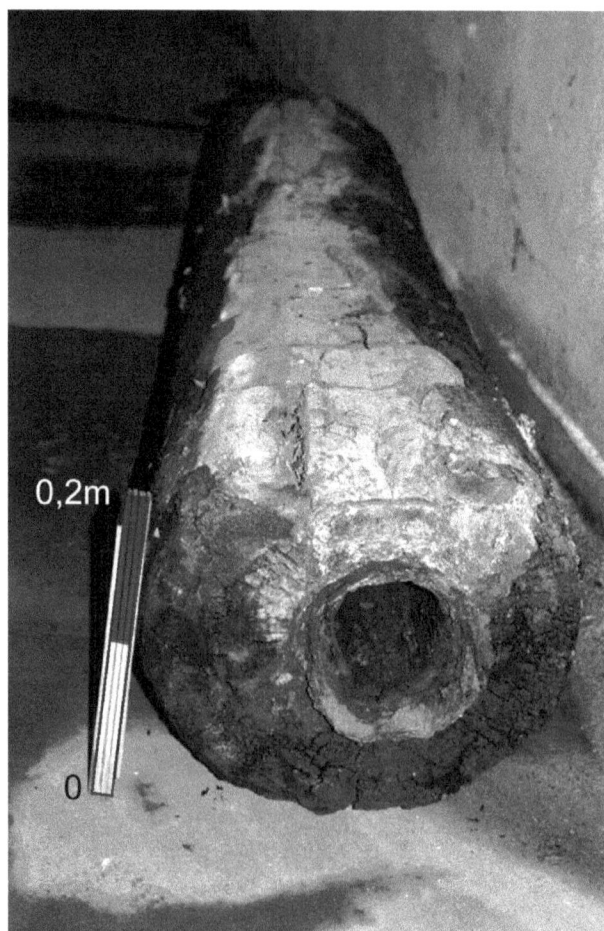

Fig. 12. Two fragment of the wooden water pipe from the excavations of Gold Street at Kalisz (fot. L. Ziąbka).

Preserved written sources pertaining to the history of medieval Kalisz, namely: 1) ducal and royal documents from between 1264 and the end of the 15th century; and 2) books of the court of Kalisz castle dating from 1419 on, enable scholars to reconstruct no more than a few ways in which this town acquired water. Moreover, the documents refer only to the "*extra muros*" space, i.e. to the suburbs, which constituted – as in other medieval centres – the town's direct food and production base. An analysis of the Kalisz document archives, done many years ago by an outstanding researcher Marta Młynarska, showed that the fact that Kalisz was situated in the richly watered Prosna river valley, was of crucial importance for the functioning of numerous watermills in the town (MŁYNARSKA, 1960, I, 111; KUCHARSKI 2004, 29-43). Archival research in the above-mentioned books of the court of Kalisz castle supported this thesis, providing information about the temporal framework of its functioning and situation in the suburban space (for instance, In 1394 the mill in the Old Town of Kalisz was mentioned, with the road leading through the property of the vicar from Dobrzec; the use of the road required special permission of the Kalisz judge) (SENKOWSKI, SUŁKOWSKA, 1960, 309 - n. 30, orig.: AGAD n. 3396, printed: KDW III, n. 1955). Providing a bit of guidance on time for its functioning, as well as on location in the

All these types of water mills contributed substantially to the economic development of the town; therefore it was necessary for them to operate constantly. This could only be ensured by well-functioning water installations of the mills, built either on the Prosna river, or on its natural or artificial branches (including leats). In the case of the plain, slow-Prosna River, it was necessary to use overshot wheels, twice as efficient as undershot ones. This meant the necessity of backwater from 2.5 to 3-4 meters, therefore, according to Professor M. Dembinska, „ the capacity to build different types of reservoirs, lakes, ponds, of different water levels, dames and dykes, channels and ditches regulating the inflow of water (DEMBIŃSKA 1973, 99). The present archaeological methods, including non-invasive archaeology, could provide opportunities to discover the remains of such constructions.

Good watering of the suburban valley terrains of Kalisz, thanks to the proximity of the river and its branches, and to the fact that the water-bearing layer occurred at a shallow depth, created favourable conditions also for the development of gardening. This fact is proven by numerous mentions about suburban gardens in the books of the court of Kalisz castle and in town court books (e.g. *domus et hortus - - in suburbio Calisiensi - - penes torrentem versus religiosos Bernardinos[1]*). Unfortunately, these mentions do not give relevant information – known from other centres – about the ways in which the gardens were supplied with water. Therefore, there is no evidence either that water was being carried from the river or that there existed dug wells in the gardens, or canals/ditches dividing the gardens. But such situation of the garden of Kalisz was close to the theoretical idea as described by Piero di Crescenzi *(Ruralium Commodorum libri XII)* in the early 14th century: „a garden needs sources or a running water course, which could be, if necessary, directed to the beds, or a pond nearby, or at least a well to take water from, or, to cool the plants, humidity that water brings" [2] And further on, "indeed, the garden situated in a moderate climate and watered by a running course will hardly ever trouble and will not need any care about seeding."The situation of the suburban gardens of Kalisz in the Prosna valley could rather result in the risk of their flooding. Crescenzi advised that for such humid gardens "dykes should be dug in November, to discharge the excess water" – and thus prepare the soil to the spring crop.

When it comes to the written sources concerning water supply in the space *"intra muros"* of the medieval Kalisz, the situation is even worse. A disastrous fire which destroyed the town in the summer of 1537[3] also destroyed the Kalisz town court books: the town councillors' books, and books of the tribunal of *advocatus* – the basic sources used to reconstruct the topography, and the ownership and social structure of each centre. Nevertheless, preserved books of that kind, dating from 1537 on, contain very interesting pieces of information about two ways of acquiring water. The first one, unseen in any other Polish centre, occurred after the 1537 fire and consisted in an obligatory construction of private underground rainwater canals, which were to be used as extinguishing canals, on burghers' plots [4]. The second way was allocating water from the town pipe water supplies by the town council. The oldest mention, dating

from 1540, about pipe water supplies in Kalisz was found in the town councillors' book and shows this system as a working one, under development at that time [5]. The councillors had a pipe water supply constructed in Świńska Street (*platea peccorum*) at the town's expense and at the request of a townswoman called Małgorzata: it was situated in front of her house and for her personal use. Małgorzata and her heirs were to pay the town 30 groschen (1 Hungarian florin) a year as long as they used this fragment of the pipe water supply. The conservation and repair costs were to be covered by the town council[6].

Three years later, in 1543, owners of other houses in the same street expressed their will to use the pipe water supply built for Małgorzata and her family. Four notable Kalisz burghers appeared before the same council[7] and – on their own and their neighbours' behalf – undertook to pay one groschen quarterly. One of the burghers was to collect the payments and then hand them over to the men in charge of the town money (*dominibus dispensatoribus*). A week earlier, on 18 April 1543, two Jews, Daniel and Jakub from Sieradz, inhabitants of Kalisz, probably qahal elders (one of the largest qahals in Polish towns), in an identical way as the above-mentioned burghers, i.e. on their own and their Jewish neighbours' behalf, undertook to pay 4 florins a year (one florin quarterly) for water. For this payment, the town councillors were to allow them to have water drawn from a public pipe water supply container situated in the Market Square to a Jewish street leading away from the Market Square[8], perhaps to a pipe water supply container placed in the Jewish quarter – as it was ordered six years later in Poznań – although it is also possible that it was drawn to provide water for the community Mikveh.

As far as the situation in Poznań is concerned, on 7 August 1549 its City Council (MAISEL (ed.), 1966 n. 100, p. 33), in order to avoid repeating quarrels of Christians and Jews concerning the use of the aqueduct sump on the market, decided to build two water reservoirs on the street inhabited by Jews. One was to be situated opposite the Synagogue, on Sukiennicza Street, the other – on Ciasna Street, opposite the house of Chaim the Jew, by house of Moses and Shloma. Due to significant costs of water distribution to these reservoirs for the city, the counsellors decided that the elder of the Poznan Jewish district Gould pay on its behalf annual rent (by each Pentecost) in the amount of 20 *grzywna* units (i.e. 32 florins – 16 florins per reservoir) as long as municipal water will be distributed to the reservoirs by means of two pipelines. The counsellors also decided that they or their descendants will maintain both the pipelines and reservoirs so that the municipal water is clean and healthy. Jews and Christians alike could take the water from the reservoirs, to avoid conflicts; the City Council decided that the first to come is the first to take water, regardless of their religious affinity, on pain of municipal penalties. It is hard to determine whether such installations were indeed constructed.

Fig. 13. View of the model of Kalisz-city, with the most important buildings (on the left synagogue)

What certainly was realised, though, was the Kalisz undertaking of water distribution to the Jewish neighbourhood. This conclusion is supported by the discovery of the remains of two wooden pipes by archaeologists from the Kalisz Museum in 2004 (L. ZIĄBKA 2004), at Jewish Street, also known as Golden Street.

In 2004 rescue excavations were conducted in the new medieval city within the former Jewish quarter. Gold Street, where the survey was conducted, was once at its centre. For a time the street wore the name Jewish street. At its end was one of the urban bastions. Gold Street is known in the sources from the fourteenth/fifteenth century, and often mentioned in the documents of the modern (seventeenth/eighteenth century) It was one of the oldest streets of Kalisz, also because it emerged from one of the corners of the market place.).

In two small trenches at a depth of 110-150 cm from the surface of today's street were discovered water pipes fragments of oak, with a diameter of about 26 cm, the internal hole of 10 cm and a diameter of 30 cm and an internal hole of 12 cm, repaired with wood of pine. It should be added, that these were typical dimensions, often discovered in other centres too (SOWINA 2009, 304), which confirms that Jews were vested with the usual amount of water. These pipes come probably from the late sixteenth and seventeenth centuries. Under the remains of wooden pipes, there was the layer of the

organic remains in which the fragments of pottery dating from the fourteenth-fifteenth century were identified. One of the discovered parts of water supply ran obliquely in relation to the course of the street – towards the former synagogue.

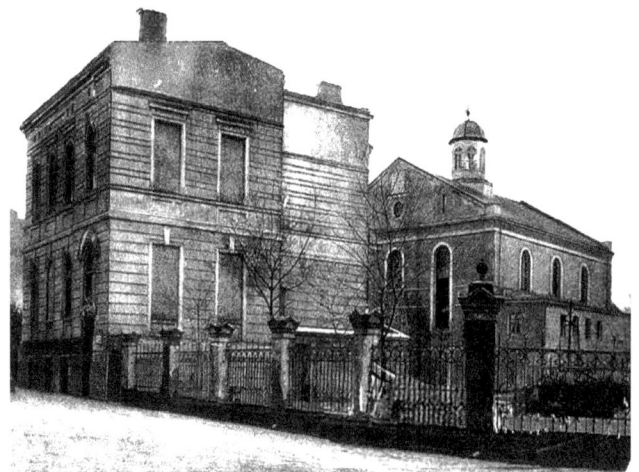

Fig. 14. Synagogue in Kalisz.

This information confirms the supposition that water was supplied in the whereabouts of the synagogue. It would be hard to determine, not having conducted further archaeological analyses, whether the discovered pipes served – as they did in Poznań – as a part of the public water supply reservoir near the synagogue, where water for everyday use was taken, or for Mikveh, which could be situated either by the synagogue, or in its basement. Only archaeological discovery of the Mikveh with parallel determining of the way of water supply could clear out these doubts. In the nearby Poznań the 15th century Mikveh was a reservoir dug to the aquiferous layer, as indicated in the contract for its construction of 1464 (SOWINA 2009, 222-225). Such a construction did not need water from the water supply network.

The Kalisz case is a rare occasion in the Polish research, where the written sources might be confronted with specific archaeological evidence.

Taking into consideration the hydrographical situation of the medieval Kalisz, one might suspect that Kalisz belonged to the few of Polish cities which did not have much trouble with water. Its situation might be compared to that of Wrocław or Strasbourg, both having shallow aquiferous layer of potable water. On the other hand, complete destructions of the city by fires over the ages, till modern times, indicate, that water management, including in emergency cases, left much to be desired. As a result, the situation of Kalisz in this respect was similar to that of cities of the same size which suffered from lack of water, such as Płock of Sandomierz, situated on high embankment over the Vistula River.

Bibliography:

BARANOWSKI T, (ed.) 1998
Kalisz Wczesnośredniowoeczny, Kalisz

BARANOWSKI T., 2004
The Stronghold in Kalisz, [in:] Urbańczyk P. (ed.), Polish Lands at the Turn of the First and the Second Millennia, Warsaw, 285-304.

BARANOWSKI T. 2010
Kalisz as an example of development of a mediewal centre in Poland, in: Buko A., McCarthy M. (ed.), Making a Medieval Town, Warszawa, 167-178.

DĄBROWSKI K., 1976
Kalisz between the Tenth and Thirteenth centuries, [in:] Medieval settlements: Continuity and Change, Sawyer D.H. (ed.), London, 265-273.

DĄBROWSKI K, 1977
Kalisz od zarania dziejów do wczesnego średniowiecza, [in:] Rusiński W. (ed.), Dzieje Kalisza, Poznań, 17-44.

DEMBIŃSKA M., 1973
Przetwórstwo zbożowe w Polsce średniowiecznej (X-XIV wiek), Wrocław, Warszawa, Kraków, Gdańsk.

GRODECKI R.,1969
Dzieje Żydów w Polsce do końca XIV w., [in:] Grodecki R., Polska piastowska, Warszawa, 595 n.

JUREK T. 1992/1993
Żydzi w późnośredniowiecznym Kaliszu, Rocznik Kaliski, vol. XXIV, 29-53.

KĘDZIERSKI A.,2010
Mennica denarów krzyżowych księcia Zbigniewa w Kaliszu, [in:] Suchodolski S., M. Zawadzki M. (ed.), Od Kalisii do Kalisza. Skarby doliny Prosny. Katalog wystawy. Zamek Królewski w Warszawie. 30 kwietnia 2010-30 maja 2010, , Warszawa, 61-68.

KUCHARSKI G. 2004
Powstanie kościoła NMP w Kaliszu i podział miasta na dwie parafie w 1303 r., [in:] G. Kucharski, J. Plota (eds), Kolegiata kaliska na przestrzeni wieków (1303-2003), Kalisz, 29-43.

PRZYBYŁ M. 2002
Mieszko III Stary, Poznań.

MŁYNARSKA M., 1960
Proces lokacji Kalisza w XIII i w pierwszej połowie XIV w. [in:] Gieysztor A. (ed.), Osiemnaście wieków Kalisza. Studia i materiały do dziejów miasta Kalisza i regionu kaliskiego, Kalisz, vol. I, 103-130.

RUSIŃSKI W. (ed.), 1977
Dzieje Kalisza, Poznań

SENKOWSKI J., SUŁKOWSKA I., 1960
Archiwum dokumentowe miasta Kalisza, [in:] Gieysztor A. (ed.), Osiemnaście wieków Kalisza. Studia i materiały do dziejów miasta Kalisza i regionu kaliskiego, Kalisz, vol. I, 293-361.

SOWINA U., 2009
Woda i ludzie w mieście późnośredniowiecznym i wczesnonowożytnym : ziemie polskie z Europą w tle, Warszawa.

SUCHODOLSKI S.,1973
Mennictwo polskie w XI i XII wieku, Wrocław.

TOMALA J., 2004
Kalisz - miasto lokacyjne w XIII-XVIII wieku, Kalisz.

TOPOLSKI J., MODELSKI K. (eds) 1999
Żydzi w Wielkopolsce na przestrzeni dziejów, Poznań.

WĘDZKI A., 1977
Kalisz w państwie wczesnopiastowskim i w okresie rozbicia feudalnego, [in:] Rusiński W. (ed.), 1977
Dzieje Kalisza, Poznań. 44-62.

MAISEL W. (ed.), 1966
Wilkierze Poznańskie, Part I – Administracja i sądownictwo, [in:] Kaczmarczyk Z. (ed.), Starodawne Prawa Polskiego Pomniki, Serie II, Part III: Prawo miejskie, , vol. III/1, Wrocław-Warszawa-Kraków.

WITKOWSKI S., 2007
Żydzi na ziemiach polskich w średniowieczu, Z uwzględnieniem Śląska i Pomorza Gdańskiego, Olkusz.

ZAREMSKA H., 2005
Żydzi w średniowiecznej Europie Środkowej: w Czechach, Polsce i na Węgrzech, Poznań.

ZIĄBKA L.,2004
Badania archeologiczne przy remoncie sieci wodno-kanalizacyjnej w ulicy Złotej nr 11, 13, 15 i skrzyżowanie z Targową, Kalisz (unpublished).

Notes:

[1] National Archives in Poznań (further: APP), Kalisz I/53, 250.

[2] Petri Crescentiensis De omnibus agriculturae partibus et de Plantarum animaliumque ; natura et utilitate lib. XII non minus philosophiae et medicinae, quam oeconomiae, agricolationis, pastionumque studiosis utiles, Basileae 1548, liber VI, p. 187.

[3] APP, Kalisz I/6: Liber pretorialis civitatis Calisiensis, p. 1 (anno 1537).

[4] APP, Kalisz I/6, p. 8 and p. 252.

[5] APP, Kalisz I/6, p. 91. Probably water was carried out from the nearby water-mill (noria?) of Korczak. Like in the other Polish towns, water supply system of Kalisz consisted of the wooden tubes (hollow tree trunks, with iron clasps) buried in the ground and of the several cisterns *intra muros*. In the year 1601 was built a second water supply system, separate, thread the Jesuit college. City employed at the time own expert called rurmistrz (German: röhrenmeister) or rurnik (Czech: rurnik), which responsibilities included building and repair of water facilities. In the eighteenth century, the waterwork has been rebuilt. In the early nineteenth century the greater part of the city were connected to the water supply system - cfr. RUSIŃSKI (ed.), 1977.

[6] APP, Kalisz I/6, p. 91.

[7] APP, Kalisz I/6, p. 177

[8] APP, Kalisz I/6, p.173: *Obligatio solvendi census per Judeos a Canalibus. Constituti personaliter -- perfidi Judei Daniel et Jacobus a Siradia incole Civitatis Calisch habentes plenum posze et mandatum ab alys Judeis incolis Civitatis Calisiensis suo et eorum nomine obligaverunt se et presentibus obligant ratione aque de canalibus manentis et profluentis florenos quattuor per integrum annum florenum unum pro quolibet quartali reponendo per triginta grossos computans et domini Consules debent illis permittere aquam de Cisterna publica ex Circulo ad plateam Judeorum quam obligationem prefati Judei facere debent apud acta domini vicepallatini. Actum in Calisch feria quarta ante festum s. Adalberti Episcopi prox. AD 1543.*

WATER SUPPLY TO THE CITY OF WARSAW (POLAND) FROM THE 14th TO THE 19th CENTURIES - THE ARCHAEOLOGICAL SOURCES

Włodzimierz Pela
Historical Museum of Warsaw

Key Words: Archaeology, Warsaw, water supply.

The oldest part of Warsaw, the Old Town, was founded on the virgin land, on the left-bank of the Vistula river valley, near the ducal stronghold built a dozen or so years earlier.

Fig. 2 The water Wheel visible in the painting entitle *Overall view of Warsaw as seen from the Praga district* by Abraham Boot, 1627.

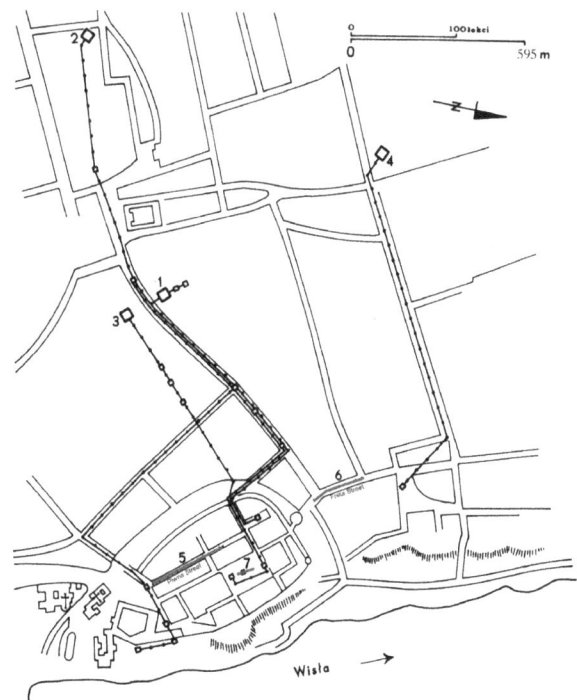

Fig. 1 Warsaw of the end of 16th century.

The founding of the town date to the period between 1294 and 1313. In the following centuries, subsequent elements of the spatial structure of Warsaw were built: New Town, built in the 15th century and the *juridices* (private towns) that surrounded it, which were founded from the 17th by the Church and gentry. The city, which as of the beginning of the 16th century became the royal residence, grew systematically and became one of the biggest cities in this part of Europe in the 19th and 20th centuries, with modern water supply and sewage systems.

Fig. 3 The water supply of old Warsaw: 1, 2, 3, 4 – Reconstruction after Plan of the Street Commission of 1773 by J. Gromski (Gromski 1977: p. 84, fig. 19); 5, 6, 7 – new archeological discoveries.

Fig. 4 The waterworks: A – wooden water pipes, Market Square, investigation 1952; B – inspection manhole with pipes visible, Wąski Dunaj Street, investigation 1999; C – wooden water pipe, Wąski Dunaj Street, investigation 1999.

Through the city's several hundred year history, the water supply problem has been analysed many times, in relation to the natural conditions of the site around Warsaw, hydrographic changes, analyses of written sources[1] (first of all bills of expenditures on the construction and maintenance of water supply devices and old technical literature),[2] as well as cartographic and iconographic data. These analyses were often of a broader character and concerned themselves with the city's sanitary culture and economy (BALCERZAK, 1968; GROMSKI, 1977).

An important role in the solution of this problem was also played by archaeological discoveries. Reports of finds of wooden pipes and wells were already noted in the 19th century (WEJNERT, 1853; WEJNERT, 1854). The 20th and 21st centuries provide further scientifically documented and varied information on the methods of supplying water to Warsaw (SZWANKOWSKA, 1953). It should be remembered, however, that in most cases, the archaeological observations took place during construction efforts and were conducted within construction excavations without the possibility of broadening their scope. This greatly limited their scope. Despite this, the archaeological materials obtained make it possible to verify extant hypotheses and observations, and provide novel, and often quite detailed, information

relating to the layout and construction of waterworks in the city.

On the basis of hydrological and historical data it is thought that at the beginning (14th-15th cent.) natural water sources were used (rivulets and the Vistula) as well as wells (CZARNECKA 1963). From the mid 16th century, information appears relating to the supply of water with wooden waterworks to the Old and New towns, and then into the castle from sources located outside of it (retention cisterns). In the 17th century, it was attempted to supply civic waterworks with water from the river Vistula below the city via a pump actuated by a water wheel (Fig. 2). The earliest layout of the water supply system preserved in documents dates to the 18th century (Fig. 3). The conduit supplying the Old Town drew water from sources located outside of the city. Using the difference in elevation and the power of gravity, it supplied it with wooden pipes into public wells found, amongst other locations, in the Old Town market square.

In addition to the waterworks supplying water to the oldest portions of Warsaw, as the city grew local supply systems were installed in noble residences. At the same time, many public wells were built in the streets and squares, as well as private wells beside homes, estates and palaces.[3] This situation didn't change until the mid 19th century (1855) due to the construction of the Henry Marconi water supply, which drew water from the Vistula and delivered it via iron pipes, and then following the construction of the modern water supply according to the design of William Lindley in the 1880's and the first half of the 20th cent.

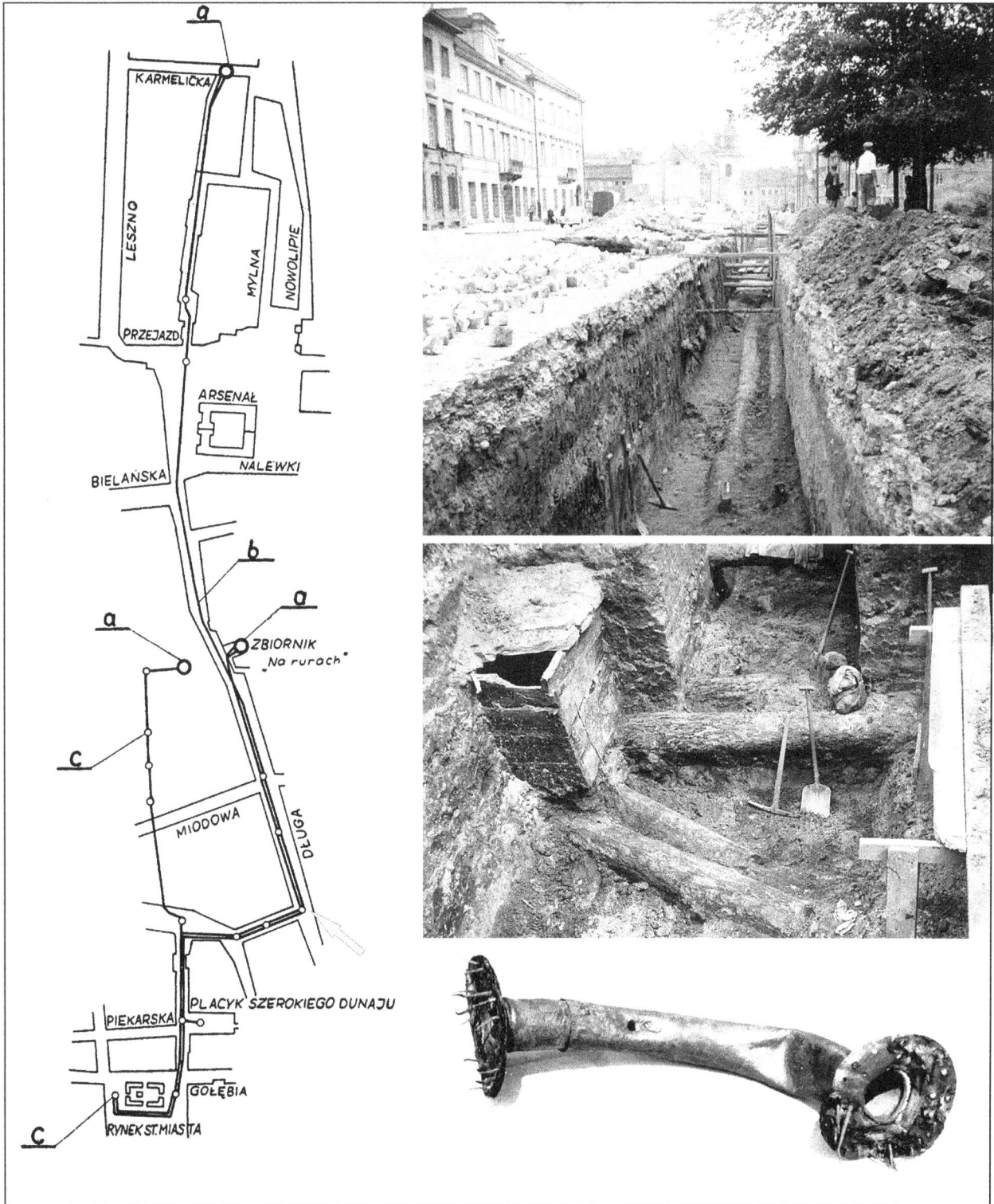

Fig. 5 The waterworks discovered under Długa Street, 1958.

Fig. 6 Warsaw, the Krasiński Square. The public wooden well: A – The wooden well visible in the painting by Bernardo Belloto, 1778; B – The wooden well, investigation 2001.

Archaeological discoveries make a few more precise determinations possible (Fig. 4). Although not much has been uncovered in terms of the precise location of watercourses and the location of the oldest wells from the 14th-15th centuries, but our knowledge of waterworks has increased, whose remains were unearthed during construction efforts. A double line of wooden pipes has been discovered in Długa and Wąski Dunaj streets. Recently, during the restoration of one of the buildings at number 28 Długa street, a wooden pipe was discovered which most likely led in the direction of a water source. A portion of the pipes discovered were not included in the water supply system from the plan drawn in the 18th cent. This pertains to both the Old Town, where a long stretch of piping was discovered under Piwna street,

likely for supplying water to the castle, as well as in the middle of the town square, where a wooden pipe was discovered at a considerable depth. This possessed an aperture for filling a well located above it. Likewise, in the New Town, a significant length of pipe was discovered under Freta street, and a wooden well was discovered under Mostowa street.

In general, it may be stated that water conduits were made of pine or fir timbers. Pipes made of them were from 1.3 to 11 m long and had a diameter of 30-40 cm. Inside the timbers, the bore had a diameter of 10-12 cm. The pipes were joined with iron joiners hammered into the bore. Where the pipe turned, metal elbows were used that terminated in metal flanges. To seal the joint, leather gaskets were inserted between the wood and the metal flange. Along the length of the conduit, access wells were placed for controlling the flow of water and collecting water throughout the water supply system (Fig. 5). The timber for the pipes discovered in 1997 was felled only in 1765, 1767, 1768 and 1770.

Wooden pipes would definitely have been replaced and various sections of the water supplies system were rebuilt, sometimes leaving older, worn out elements in place. This is evidenced by such finds as an access well discovered in 1958, in which underneath newer pipes, older piping was left behind. Unfortunately, at that time, tree-ring dating studies were not being performed in Poland yet[4], thus it will remain a mystery what the chronological relationship is between the individual pipes, and whether it would be possible to estimate the dating of the lower piping to the late medieval times, as those researchers did.

Wooden water supply piping have also been found throughout Warsaw. Unfortunately, only a portion thereof has been properly localised and archaeologically documented. The fragmentary nature of the finds and the lack of dendrochronological evaluation for the moment prevent a satisfactory analysis. We can only state in general terms, that these finds confirm the existence of many water supply systems using local sources.

The exception here is the residence of King Jan III Sobieski in Wilanów, near the borders of Warsaw of the time. During archaeological excavations performed there in the years 2007-2009, a wooden waterworks were discovered there. On the basis of written material and archaeological finds, it is assumed that it was built towards the end of the 17th century and used until the middle of the 18th century. A 120 m span of wooden pipes were found. The waterworks began within the royal grounds, at the shore of a lake where there was a spring. Here, the water was accumulated and sent under pressure to the palace gardens, where it fed two fountains. Near each of these there was a drawing well. The waterworks were built of wooden pipes with a diameter of 30-40 cm, reinforced at their ends with iron hoops and joined with lead connectors.

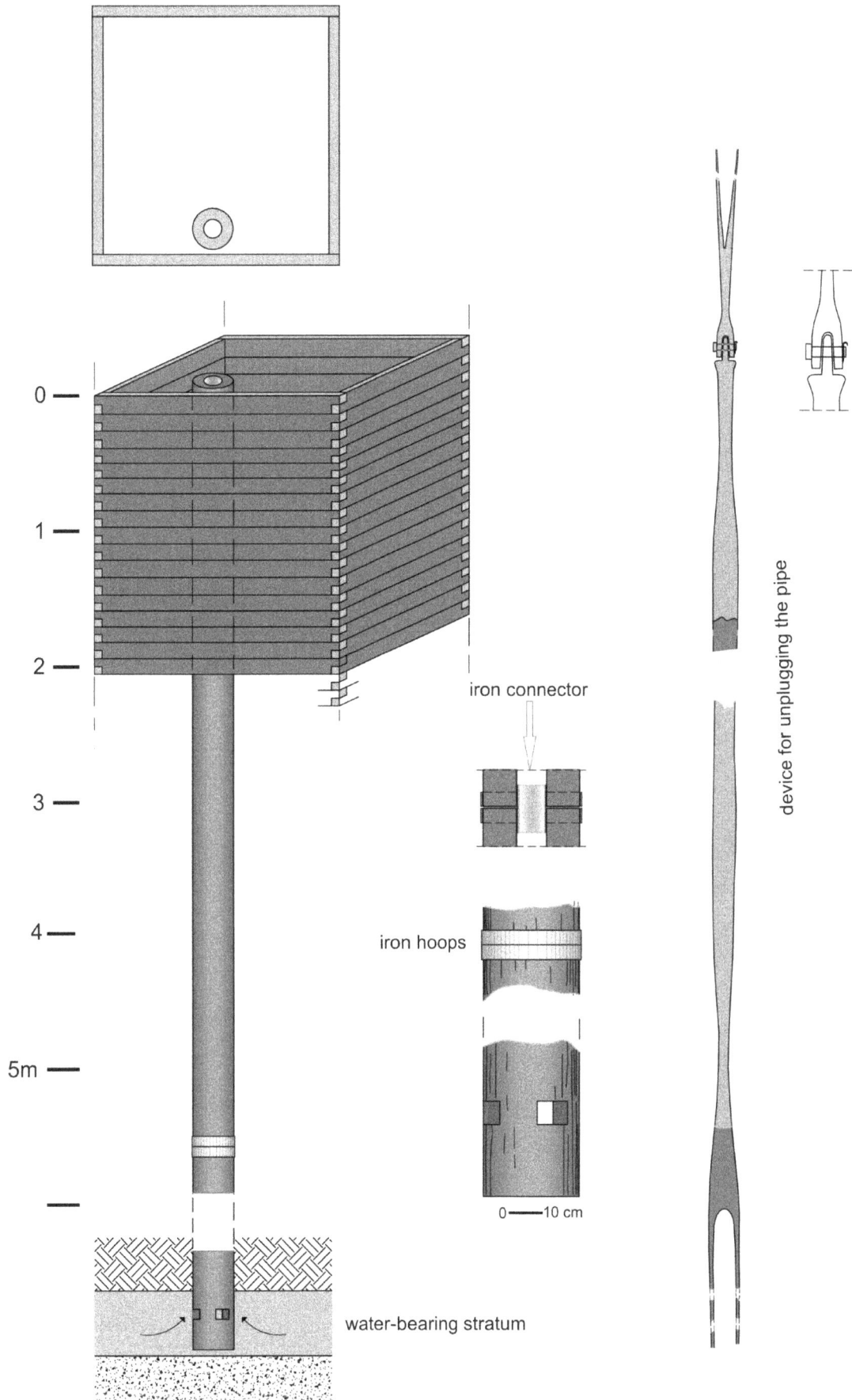

iron connector

iron hoops

0 ——— 10 cm

device for unplugging the pipe

water-bearing stratum

0
1
2
3
4
5m

Fig. 7 Warsaw, so-called Hoover square. The elements of wooden well discovered during excavation 2007.

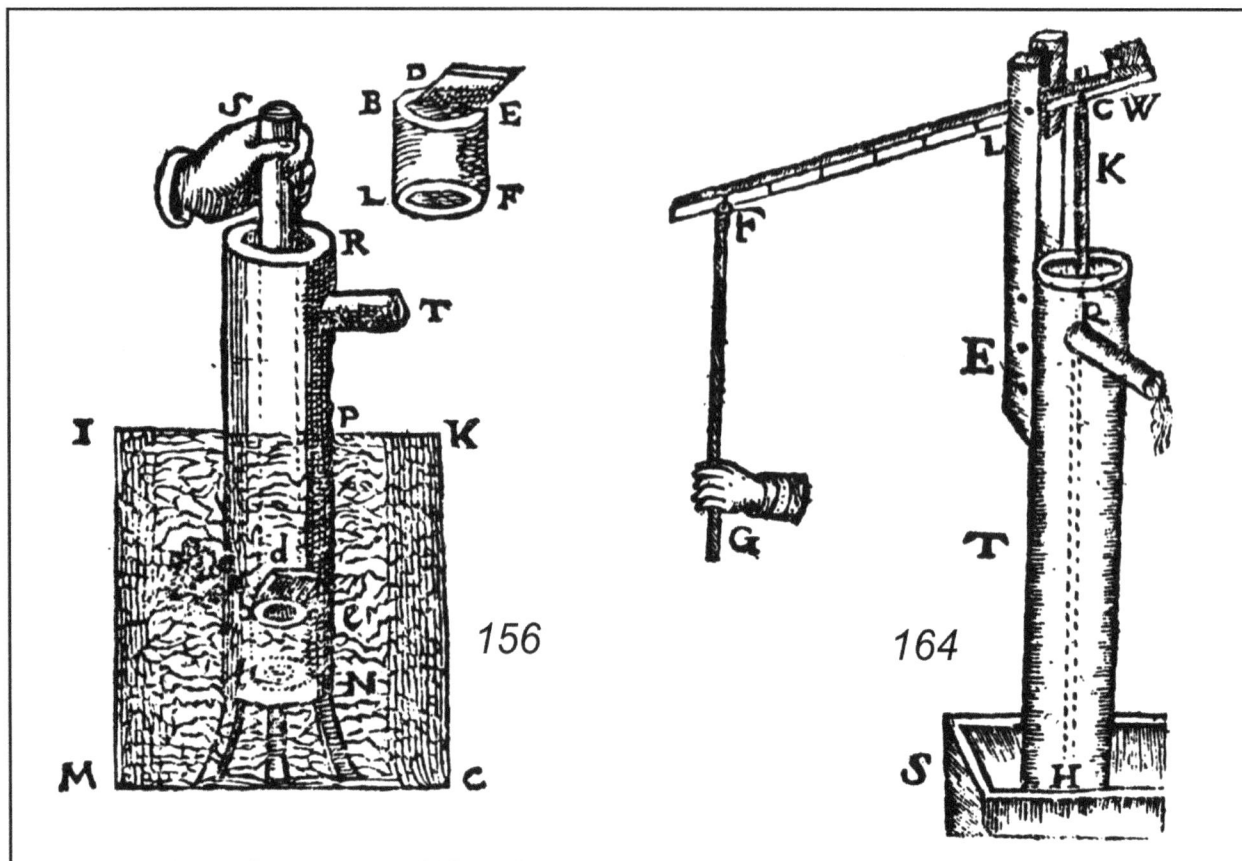

Fig. 8 Pressure pumps used in wells in the 17th century, after A. Solski, Architekt polski, fig. 156 and 164.

As mentioned above, an important part of the water supply consisted of wells, both public and on privately owned plots. They are mentioned primarily in written documents, but very few have been found in archaeological efforts. One is a wooden well from the 16th century, found in the basement of one of the oldest buildings by Old Town Square. The well was discovered during the reconstruction of the building in 1913. It was found some 10 m below the present level of the square, in a masonry-lined chamber then interpreted as a dungeon.

From later times, the analysis of a well from the Krasiński Palace is quite interesting. The upper, aboveground appearance of the well is known from a painting by Bernardo Belloto *vel* Canaletto ("The Krasinski Palace" – 1778), whereas the belowground portion was unearthed and documented during construction efforts in 2001 (Fig. 6).

Among many remains of wooden wells discovered during excavations, the structures dating to the 18th-19th centuries are very interesting. Their construction made use of the hydrographic structure of the Warsaw embankment. These objects, in addition to a wooden structure (crate), had a wooden structure placed inside consisting of wooden pipes joined with iron connectors, additionally reinforced at the ends with iron hoops. The lowermost pipe, reaching the water-bearing stratum, was not tapped all the way through and above its bottom were perforations permitting the flow of water forced upwards under pressure. Inside the pipe there were devices for

unplugging the pipe. This type of solution was found in several spots, and the most interesting was found in the so-called Hoover square where in addition to the structural details of the well it was possible to determine their location within the confines of a private plot (Fig. 7). It is very likely that they constituted a part of the pressure pumps that functioned based on the principles described by A. Solski in the 17th century. (Fig. 8).

To summarise, it seems that the role of archaeological sources in the explanation of questions relating to the water supply of Warsaw is important, but very uneven and fragmentary; both in relation to various chronological and spatial periods throughout the ages of the city. It is also significant that due to detailed data, archaeological investigations make it possible to illustrate information obtained from other historical sources. Archaeological sources can also be a valuable inspiration for other queries in the archives, particularly for the modern era (18th – 19th cent.)

Bibliography:

BALCERZAK, E., 1968
Zaopatrzenie w wodę miast mazowieckich w drugiej połowie XVIII wieku, Wrocław.

CZARNECKA, H., 1963
Źródła na terenie Warszawy, *Wiadomości Służby Hydrologicznej i Meteorologicznej*, no 54a (2a/1963), pp. 3-21.

GROMSKI, J., 1977
Kultura sanitarna Warszawy do końca XVIII w., Warszawa .

SOLSKI, A., 1959
Architekt polski, to jest nauka ulżenia wszystkich ciężarów... , Wrocław (the first edition - Kraków 1690).
SZWANKOWSKA, H., 1953
O wodociągach Starego i Nowego Miasta, Ochrona Zabytków, 8, no 2-3, pp. 128-131.

WEJNERT, A., 1853
O wykrytych dawnych wodociągach i wykopalisku przy Kolumnie Zygmunta III, *Tygodnik Ilustrowany*, no 43/44.

WEJNERT, A., 1854
Odkrycie wodociągów dawnych Nowej Warszawy, Warszawa.

Notes:

[1] The oldest known sources originate from 1482. (Protocols in Old Warsaw council books)

[2] Particularly noteworthy is S. Solski, *Architekt polski*, who gives descriptions and reconstructions of many methods of drawing water, lifts and pumps.

[3] Based on preserved descriptions, it is known that in 1796 there were 25 public wells and 1988 private ones in Warsaw (98.7% private). This amounted to one well per over 2000 citizens (the population of Warsaw (minus the eastern shore population of Praga) was 57294 at the time (BALCERZAK, 1968, 101).

[4] Attempts to date the wood found during archaeological excavations were undertaken only in the mid 1960's.

www.ingramcontent.com/pod-product-compliance
Lightning Source LLC
Chambersburg PA
CBHW061005030426

42334CB00033B/3372